Testimony for Earth

A Worldview to Save the Planet and Ourselves

"After the Storm — Kootenay Lake"
by Maggie Oliver

Praise for *Testimony for Earth:*

Everyone should read Harringtons' *Testimony for Earth*. This book is a compelling guide and a call to action for everyone to make necessary changes to restore and protect the Earth. Bob combines his experience as a conservationist, philosopher and educator to give us well-thought-out solutions to make our planet healthy again. Our life values must also change to include respect for all things, live more deeply, and recognize we are connected to everything on Earth.

COLLEEN MCCRORY, WINNER:
1992 UN GLOBAL 500 ROLL OF HONOUR
1992 GOLDMAN ENVIRONMENTAL PRIZE FOR NORTH AMERICA
1993 GOVERNOR-GENERAL'S CONSERVATION AWARD

Long time naturalist Bob Harrington joins the tiny handful of deep green ecocentric writers who offer an alternative to the destructive ideology of our age — the inward looking, unshakable worldview that our species is more important than anything else on the planet and that we have a right to rape and pillage nature at will. The book's foundation rests largely on the eleven principles of the recently published "Manifesto for Earth," a deepest green secular worldview born of ecological understanding of humanity's place in nature. The author brings his own stories and a personal spiritual dimension to his interpretation of the manifesto's unifying principles whose goal is the restoration of earth's diversity and beauty.

DR. TED MOSQUIN, CO-AUTHOR OF
"A MANIFESTO FOR EARTH"

Testimony for Earth

A Worldview to Save the Planet and Ourselves

by Bob Harrington
with Linda Harrington

Illustrations and cover painting by Maggie Oliver

It is testimony that is needed. By some happy chance Rosalind had hit upon the one word that had the power to activate the fighter in the old woman: testimony. A human being, to live a meaningful life, was required to bear testimony; in prayer, in the husbandry of the home, in the conduct of public life, a man or woman must at critical moments testify publicly as to fundamental beliefs. Ruth Brinton had always done so, which was why she was regarded throughout the eastern Shore as a Quaker Saint, difficult at times, stubborn always, but a testament to man's striving for a saner life.

—James Mitchener, *Chesapeake,* pg. 383

ISBN-10 0-88839-645-7
ISBN-13 978-0-88839-645-7

Cataloging in Publication Data

Harrington, Robert F.

Testimony for earth : a worldview to save the planet and ourselves / by Bob Harrington ; with Linda Harrington ; illustrations and cover painting by Maggie Oliver.

Includes bibliographical references and index.
ISBN 978-0-88839-645-7

1. Earth. 2. Environmental responsibility. 3. Environmental ethics. 4. Human ecology—Philosophy. 5. Nature—Effect of human beings on. 6. Environmental protection.
I. Harrington, Linda II. Title.

GE42.H368 2008 304.2'8 C2008-901264-X

Printed in Indonesia — TK Printing

Editor: Theresa Laviolette
Production: Mia Hancock
Cover design: Mia Hancock
Front cover: "After the Storm — Kootenay Lake," oil painting, 20 x 30 inches, Copyright Maggie Oliver 2007
Chapter illustrations: graphite sketches, 4 x 5 inches, Copyright Maggie Oliver 2007

We acknowledge the financial support of the Government of Canada through the Book Publishing Industry Development Program (BPIDP) for our publishing activities.

Published simultaneously in Canada and the United States by

HANCOCK HOUSE PUBLISHERS LTD.
19313 Zero Avenue, Surrey, B.C. Canada V3S 9R9
(604) 538-1114 Fax (604) 538-2262

HANCOCK HOUSE PUBLISHERS
1431 Harrison Avenue, Blaine, WA U.S.A. 98230-5005
(604) 538-1114 Fax (604) 538-2262

***Website:* www.hancockhouse.com**
***Email:* sales@hancockhouse.com**

Contents

Core Principles

Action Principles

DEDICATION

Dedicated to
Stan Rowe
(June 11, 1918–April 6, 2004)
and
Ted Mosquin

Both of these scientists collaborated in writing "A Manifesto for Earth," which sets forth eleven principles of life-saving importance to our jeopardized planet. These principles form the nucleus for this book.

Introduction

The Importance of "A Manifesto for Earth"

I have not found anything in life to be more meaningful than the Earth. Since boyhood I have preferred to live in small, quiet places where woods and fields are nearby. Walking has always been my preferred means of locomotion. I recall that when I was ten or twelve years old I kept a little journal that I labeled "Wanderings and Musings." The world was less noisy then. There were airplanes and cars, but it was possible to walk along roads leading to a fishing hole or a berry-picking spot without seeing an automobile more than occasionally. Even in youth I was not particularly attracted to towns and felt no inclination at all to visit any large city. I still feel little attraction for such places. I have lived in British Columbia for nearly fifty years and have never been to Vancouver or Victoria and do not feel deprived. To the very best of my ability I have tried to remain part of the natural world that I love. Somehow I believe that the subtleties of Nature communicate hints, moods, feelings and insights to our unconscious mind.

One of my strongest convictions is that humanity has unwittingly created a perilous situation by detaching itself from Nature, and in considering itself to be superior and the only species that is important to the universe. The conquest of Nature, which has been expressed as a goal by modern entrepreneurs, has always seemed to me a sign that our species is disloyal to the very source of its existence. I do understand that some theologies postulate that we are direct offspring of deity (as are all other organisms?) and have been given Earth as a gift for our private use. I suspect this is at best fanciful, and this is why I proposed in *The Soul Solution* that a theology of the Earth might elevate us to a deeper reality and a willingness to care for our planetary home.[1] It seems to me that if we believe in a creator we should also show great respect for the creation designed by that creator. Our maiming of ecosystems and the ecosphere indicates that such respect does not exist.

One of the reasons I decided to write this book was because of my acquaintance and friendship with the late Dr. Stan Rowe, professor emeritus from the University of Saskatchewan. Our mutual interests in ecology made us compatible companions and we would walk, talk, or communicate by letter or telephone. In 2004, Stan with his friend and colleague, Dr. Ted Mosquin, a specialist in systematics and evolution and a widely accomplished man, teamed up to create "A Manifesto for Earth." This was a scientific article that first appeared in *Biodiversity: Journal of Life on Earth* (January–March 2004, see appendix). Rooted in ecology, it offered eleven principles that, if followed, would move humankind toward a respectful, compatible relationship with Planet Earth and the multitude of other organisms with which we share this life. They hoped many of the most idealistic ideas society has conceived would become part of our thought if we realized our intimate ties with other members of the super-family of life on Earth.

Ted Mosquin gave an interesting perspective on the intent of the Manifesto authors when he addressed members of the Canadian Association for the Club of Rome (CACOR) in 2005. In answer to the question "Why another Manifesto?" Dr. Mosquin clarified the viewpoint adopted by Dr. Rowe and himself.

"Some reviewers of the early drafts of the Manifesto pointed out that humanity already has a half dozen charters, proclamations, platforms, declarations and the like, and suggested to the authors that we explain why yet another of these kinds of texts was needed. The answer is that all these earlier attempts are thoroughly or partly anthropocentric in that they place humans rather than the living Earth at the center of value. They reflect a deeply embedded anthropomorphism, seeing the world of Nature as being valuable only to the extent that it served some human purpose. Hence, the Manifesto was written to fill this gap in the literature on environmental ethics by explaining the ecological basis for the ecocentric worldview."[2]

Inasmuch as I also share the idea that the Earth is of vastly greater importance that any of its expendable species, I feel that an ecocentric worldview is a prerequisite for truly civilized behavior that would always keep a weather-eye alert to the health of the Earth context in which we live. In communication with Ted Mosquin I was impressed by a thought he expressed, that since we always capitalize the names of other planets — Venus, Jupiter, Neptune, for example — we should also capitalize Earth as a matter of respect. This led me to the further thought that addresses are incomplete without Earth completing the full address. Having worked as an educator, I reflect that when children learn their addresses, they should be taught that Earth is a natural part of that address. Because we display habitual unconcern about our planet, it might be a valuable part of all education to remind people at an early age that Earth is home. An international ballad, "O Earth, We Stand on Guard for Thee," might help all humans to realize that wars in all their forms — civil, religious, or international — are primarily destructive and no wiser than bashing your own feet with a sledgehammer. It is quite amazing that politicians have not yet come to the realization that peace is an inviolate need.

After reading through "A Manifesto for Earth" a number of times, and pondering upon the eleven principles stated by its authors, I recalled that some years ago Buckminster Fuller wrote a book he called *An Operation Manual for Spaceship Earth.* Fuller had lamented on more than one occasion that such a manual was not available. Realizing that operating manuals are given for everything from lawnmowers and garden tractors to refrigerators and sewing machines, it could be said that he had a valid point. Fuller's book made many pertinent observations. Earth, though, with its complex living, dying, reproducing, mutating multitudes of amazingly specialized and integrated organisms, is infinitely more than a spaceship. It is an ecocentric masterpiece that not only nurtures its offspring, but also is the parent of philosophies and spiritualities that hint of unknown destinies. It then occurred to me that the Mosquin and Rowe essay would be a realistic foundation for a manual to guide wise and respectful beings toward a harmonious, mutualistic future on the planet.

When I thought about the idea of a manual for Earth, I became conscious that although Canadian politicians may be 24-carat whizzes in political science matters, they frequently demonstrate acute estrangement from the Earth. They do not appear to realize that Canada is a vast land with an integrated storehouse of enormous vitality. The preservation of that vitality is the essential factor upon which any intelligent government ought to be focused. A minority of legislators may understand this because they have sprung from a farming background, or have comprehensive knowledge of some scientific discipline; but political science is too thin a discipline to engender knowledge about the importance of concern for the health of the planet.

There are many other people whose narrow focus overlooks concern for Planet Earth. Businesses, or corporations, are too pre-occupied with wealth to have regard for anything more important than the economy. Many voters accept the glamorous pretensions of politicians while they themselves possess only a superficial awareness of Earth as, literally and figuratively, the miracle that supports all life every day of every year. And still another important segment of society, the educational system, focuses on human achievements and aspirations and, without apparent intent, fails to root young people in the Earth context that makes their lives possible. An instructive Earth manual would be of immense value to each of the groups just mentioned. The lack of such a valuable "operating manual" is apparent today.

Upon reading the eleventh principle of the Manifesto — Spread the Message — I felt that I should do just that by providing some general expansion on the principles stated in the Manifesto and attempting to give insight into some of the problems we face.

Unfortunately our society has set so many negative factors in motion that it will take a monumental awakening to divert ourselves from the abyss toward which we are accelerating. Just as much as any religion seeks renewal, people need a revelation, both logical and spiritual, to help them become loyal citizens of our planet. This statement leads me to think of another aspect of the Manifesto. It is that the principles stated by the authors are not merely scientifically valid. Their manuscript transcends strictly academic science, as I think it must do. It displays both philosophical and spiritual conviction. It does not hesitate to speak of some behavior as "morally reprehensible." It offers hope by suggesting behavior that would open a "new and promising path toward international understanding, cooperation, stability, and peace." Focused on broad ecological concepts, it displays a truly holistic Nature.

In writing this book I am also prompted by my own deep respect for Earth. This respect probably led me to studies in geology, and later in ecology. My interest also caused me to peruse writings in classics to better understand early attitudes toward Earth. As indicated in a later chapter, concepts that eventually formulated the idea of Earth wholeness appear throughout history. Ecology is a science that had its inception in the thoughts of sages whose intellects had not disassociated physical factors from moral evaluation. Its integrative Nature and ecocentric worldview is vital to human survival and extremely necessary at this time. Sharing Mosquin's and Rowe's own holistic outlook I have tried to expand on some of the factors, concerns and attitudes that such a worldview entails.

Since it has long been known that fools step in where angels fear to tread, I can justify my own attempts to spread the message while being fully aware that I will step on some toes and perhaps on some cherished (but I think hollow) ideals, such as one espoused years ago that "the business of Canada is business."

I believe strongly that the ideas of the Manifesto are not merely useful but are direly needed. What I hope to do is to spread the message more widely so that the idea of ecology as a taproot for social structure, for the economy, for education, and for survival on Earth will become the realistic conviction of enough people to make a difference. Instead of trying to shape the world to fit our own desires, we must file down our own prominent abrasive edges and become constructive members of the ecosphere.

Our present commitment — giving the economy top priority in our lives — is steadily degrading the Earth. Actually, there are three evidence-filled concepts in our awareness, which we view in the wrong order of importance. As we think of them they are economy, education, and ecology. To understand how we have reversed the proper order of priorities, we must visualize ourselves as part of the cosmos, which scientists have described as a seamless whole. The word cosmos, from the Greek *kosmeo,* means "order" or "arrangement." An example of cosmic order is our solar system, which consists of planets regularly orbiting a star we call the Sun. Predictability is a factor in our cosmos.

The incongruity of selecting the economy as our most important concern is becoming clearly evident. The need for rethinking our own priorities is enormously important.

Two of the three concepts mentioned above share a common root, eco, derived from Greek *oikos,* which means household. In the order in which they should be emphasized, the first of these words is ecology, which is senior because it means the study of the household. It is logical that the study of the household should precede economy, which refers to the management of the household. It becomes more obvious daily that, if we knew more about our household, we would not manage it as badly as we do. Our world would not be filled with life-threatening garbage and pollution, which contaminates our soil, air, and water. Ubiquitous toxicity is a direct result of our incompetence as planetary managers.

The third term links ecology and economy. It is the word education, which means, to lead away from. In other words, the process we should follow is to carefully study the organization of our planet (ecology) and then transmit proper understanding (education) to those who are responsible for harmoniously blending our lives into the seamless whole pre-organized by the innate productive order of the cosmos (economy). Such an understanding would significantly change our behavior, which currently verges on suicidal brinkmanship.

The Message of Faust

Planetary ecology vastly antedates humanity, and is activated by the portion of the sun's energy captured by Earth. This will be explained in a later chapter. At present we can recognize that something is seriously wrong with our behavior and is a problem that has beset humans for ages. The best place to find an answer is to look back in history. In *To Heal the Earth: The Case for an Earth Ethic* (1990), I referred to Goethe's *Faust* as a wonderful analogy to help us understand the problem that drives us toward self-destruction.

The play *Faust* by Johann Wolfgang von Goethe is based on a theme that dates from antiquity, and is also descriptive of the times in which we live. Faust, the central figure of the play, in effect sells his soul to the devil in order to obtain power and knowledge.

Many aspects of modern society repeat the *Faust* theme in that humankind has been persuaded to give up its simpler and natural relations with the reality of Nature and, as Albert Schweitzer reflected, "to seek its welfare in the magic formula of some kind of economic and social witchcraft, by which the possibility of freeing itself from economic and social misery is only further removed." He was aware that once we gave up our individual personalities and succumbed to the witchcraft of economic dominance we would have to give up our own ideals and accept the materialistic philosophy of the masses.[3]

Faust, the hero of the play, is saved from damnation by his recurring realization of the integrity of Nature, and later by taking an active role in restoring land to agricultural productivity that had been flooded by the sea. In similar fashion, the salvation of people today will only be found by working wholeheartedly to restore the Earth to health. This will entail a more frugal, less machine-intensive, pollution-banning form of life. It will involve serious change in our ways, and necessitate control of corporate actions. It will not be an easy job because of the reckless and even demonic way that Nature has been ravaged. In the parlance of World War Two days, we will have to choose between "shaping up or shipping out." It is not surprising that many ordinary people realize that Nature is presenting her bill and expects payment. Win or lose, it would be a remarkable saving grace for us to finally cherish what has been nearly ruined, and thereby become spiritually healed.

About 600 BC the *Tao te Ching,* attributed to Lao Tzu, identified the universe as a sacred vessel which was not made to be altered by humans. The *Tao* advised that the world would be ruined by tinkering with its established order. Basic to the *Tao* was the idea that we are incapable of comprehending Earth's wholeness.[4]

In *Faust,* Goethe makes it clear that separation from Nature is the greatest error into which man can fall. The central message of the play is the *danger* of such alienation from the security that Nature affords. People today have successfully alienated themselves by turning their backs on reality in their pursuit of trivial distractions. In *Faust,* the hero is subjected (as we have been) to numerous erring adventures. After each of these he returns to a new life in Nature. At last, there breaks through in his heart the "longing to once more win, at whatever price, a normal relationship with Nature." This might be our saving grace. This stanza stands out in the Faust drama:

I have not yet fought through to liberty.
If I could from my path all witchcraft banish,
Let all the formulas of magic vanish,
Stood I a man before thee once again,
That would be worth, O Nature, all the pain.
A man I was before I sought the shadows.

Goethe (1749–1832) lived in a period during which the ethical way of life was considered to be promotion of the common good. The flaw in this concept was that while the majority of members might make sacrifices from which no gain would accrue to them, other individuals acting purely from self interest would attempt to gain an undeserved amount of personal prosperity for themselves. Profiteering is prevalent today. The problem with ethics demanding that everyone work for the common good has always been that those who do not participate will easily gain at the expense of others. Goethe realized that a good society is one in which respect, reverence, and a sense of duty rise *upward* out of an individual's ethical character, which stems from his regard for the infinite. Some philosophers [e.g. Adam Smith (1723–1790) and Jeremy Bentham (1748–1832)] believed altruism to be a derivative of ego. Goethe disagreed. We know today that if altruism did stem from the ego then we would not have a world dominated by extreme wealth and power. Today's critical situations would not have occurred because excess wealth would long ago have been turned actively toward eliminating the manic exploitation of our planet. Whales would not have been exploited to near extinction. Fishnets miles long would not have been used. Selective logging practices would have been followed out of respect for land, rather than clear-cutting with elimination of even tiny seedlings. Well-organized mass transportation would long ago have reduced personal vehicle use. People would long ago have been educated to the preciousness of our planet and its need for judicious use, respect, and protection from the irrational greed that now dominates society. Exploitative, planet-wrecking behavior is a direct result of unbridled ego and turning our back on the real world.

Goethe, I believe, was correct in seeing the world as a manifestation of the infinite spirit. He did not mock religions but displayed a personal certainty that God and Nature are entwined.

This, of course, makes sense if omnipresence, omnipotence, and omniscience, are attributes of God. Although more a Stoic than a Christian, Goethe punned that from his thoughts and devotion to Nature, he "might be the only Christian."

Led by economic single vision, humanity is in a state of suicidal rebellion against the natural order from which it emerged. The goal of conquest of Nature is an aberration. Control of selves and of the economic fantasia we are seeking should focus us on becoming loyal citizens of Earth.

In *Generation of Vipers,* Philip Wylie observed, "Anarchy exists nowhere in Nature. An asceticism, which is to say, a discipline, is imposed upon every living thing by its environment and its instincts...Common man has at long last got himself so far out of gear with Nature and his environment that he is beginning to see the shape of extinction, whether he likes it or not."[5]

Recognizing and Facing a Crisis

Concrete evidence of humanity's sorry plight was presented in the January 2007 issue of *ecologist*. Zak Goldsmith's editorial quotes International Red Cross Annual Disaster Reports, which reveal that the number of people affected by natural disasters increased from 275,000 in the 1970s to 18,000,000 in the 1990s, a sixty-five-fold increase.[6]

UNEP director Klaus Toepfer contends that the number of people seeking to escape creeping environmental destruction by 2010 will be fifty million. The Intergovernmental Panel on Climate Change (IPCC) envisions 150 million people seeking such escape by 2050. The Stern Report suggests that 200 million people will be permanently displaced from homes by rising sea levels caused by a temperature increase of two degrees. (The present average temperature increase worldwide is 0.7 degrees.) Factors contributing to such projections as these include the shrinking of the Greenland icecap by eleven cubic miles per year, and The World Glacial Authority report that seventy-five of eighty-eight glaciers it has studied are shrinking. Also on the dry side, China's Gobi desert is growing at the rate of 10,000 square kilometers per year.[7]

The report on economic impact of climate change by Sir Nicholas Stern, former World Bank economist, recruited for that effort by England's chancellor Gordon Brown, was completed on October 30, 2005. Its message was grim, and stated that we are in a race against time to avert climate catastrophe. The *ecologist* states that Stern's "core message is inescapable: the end of the world is nigh." His conclusion is that the science of climate change is incontrovertible and we may ignore it only at our own peril. He refers to the business as usual concept as the "economics of genocide."[8]

As part of the *ecologist* presentation on climate change, Anna da Costa's report countered Stern's attempt to give some economic comfort and a few years time to producers. Her presentation included a sequence of images made available by the Hadley Centre for Climate Change, which show that unless large reductions in emissions are quickly achieved, parts of the Earth will be experiencing temperatures three degrees above pre-industrial levels by 2020. The images have not previously been made public except at "climate clinics" for MPs and their friends. They are grim. The report by the IPCC in 2001 that global warming is the result of human activity was verified by the Stern report and although Stern might like to engineer change in such manner as to keep impact on the GNP to a minimum, this would be an extremely dangerous risk. The current level of greenhouse gases (GHG) in the environment is 430ppm (parts per million) and is increasing 2.3ppm per year. Even with vast efforts, stabilization at 450ppm is nearly out of reach.

Not only the Hadley Centre but also the Tyndall Centre for Climate Change Research urged recognition of the seriousness of today's problem. The Hadley Centre projections were kept from public knowledge "simply because of their unpalatability, and because of the frightening message they contain." Realizing that the material I am attempting to encapsulate is very lengthy, I will state briefly what the researchers recommend. The Tyndall Centre states concisely that we cannot afford to wait for technological innovations (which may never come) or for political schemes such as carbon trading. The Tyndall Centre claims "we need worldwide to reduce our carbon emissions by an unprecedented nine percent a year for up to 20 years."[9] What we are heading for if we stay on our present path is "a scorched Earth right across the planet." If this book has a purpose it is as a wake-up call for humankind.

It is obviously time for us to subordinate our behavior to the imperative requirements of Planet Earth. Governments and citizens must realize that changed behavior is not a political issue but a moral issue, as well as an ecological ultimatum.

[1] Harrington, Robert F. *The Soul Solution* (British Columbia: White Oak Press, 2000) pp.174–190.
[2] Mosquin, Ted. "Some Thoughts on the Manifesto for Earth." Proceedings: Analysis of the Human Predicament, May 2006, p.31.
[3] Schweitzer, Dr. Albert. Quotation from a speech given as the centennial celebration of Goethe's death, in his native city, Frankfort on the Main, March 22, 1932.
[4] *The Soul Solution,* p.103.
[5] Wylie, Philip. *Generation of Vipers* (New York: Holt, Rinehart and Winston, 1942) pp. 103–11.
[6] *ecologist*, London, England, January 2007, p.005.
[7] Ibid. p.005.
[8] Ibid. p.010.
[9] Ibid. pp.010–014.

1. The First Core Principle • The Ecosphere is the Center of Value for Humanity

Planet Earth is our home. It is the core of the ecosphere, which supports all organisms. A likely question at this point is, "What is the ecosphere?" Think of it as the whole round world complete with its atmosphere and its protective ozone layer that filters out some of the more harmful frequencies of sunlight. It includes the oceans that make up three-quarters of Earth's surface and the deserts, which are growing as a result of human activities. It includes the fertile land surface with its grassy meadows, the ravaged remnants of its majestic forests, its lakes, streams, soaring mountains, glaciers and all other special features that make Earth so precious. Countless vistas speak to our awareness and are rightfully entitled to be cherished. Who are we? We are parts of the miraculous web of life — the interplay of organic and inorganic processes entwined in mazes of complex relationships.

The idea of relationship among all living things goes back millennia. People long ago spoke of Father Sun and Mother Earth. Today, I read that there are ten thousand religions and many of them claim that our own species is the only one of importance. This belief is not only homocentric, but exclusively focused on people who share membership in what is denoted the only correct creed. I am reminded of a comment that claimed that if cows believed in God, He would look like a cow. Other religions are more charitable, and the Buddhist faith has been described as biocentric in its outlook. This means valuing all life. Scientists have reminded us that in spite of the exalted position we sometimes assume, we are simply food to other organisms such as fleas, bedbugs, ticks, mosquitoes, black flies, and numerous other creatures.

What is Life?

We might understand our relationship to the universe better by reflecting upon a vivid definition of life given by Winwood Reade, in 1872: "Glorious Apollo is the parent of us all. Animal heat is solar heat; a blush is a stray sunbeam; life is bottled sunshine, and Death the silent-footed butler who draws out the cork."[1]

Reade, an Englishman, explored the headwaters of the Niger River in Africa. Perhaps his exposure to sunlight and heat in Africa led him to such an illustrative clarification of the age-old question, what is life?

The foregoing analogy of life is, I feel, instructively vivid. In a telephone conversation with J. Stan Rowe, who, with Ted Mosquin, wrote "A Manifesto for Earth," Stan and I spoke of the lameness of many of the explanations of life. When I later included Reade's quote in a letter to him, his reply was enthusiastic:

"Bob, I love Winwood Reade's definition of life, (as bottled sunshine). Of course as the cork is drawn, new sunshine pours in. The advantage of adopting the metaphor, Earth = Life, is its twin, Earth = Death, two sides of the same round coin, merged in recurring cycles. Without Death no Life, without Life no Death."

Stan went on to state that he didn't have the "temerity to explain life." One of the purposes of the Manifesto, he stated, is to encourage people to disengage the idea of life from organisms, from things like us. He felt that if people understood "that Sunny Earth is the life-giver, the life-carrier (rather than organisms = life) then we might treat the planet better, and at the same time get off the narcissistic kick that positions us as the most important thing in the universe. Earth still has four billion years to produce the intelligence so far lacking, unless we ignoramuses foreclose the potentialities in the next 50 years or less."

It is difficult to understand why humans need to feel superior to other life forms. Certainly microbial organisms of various kinds have performed amazing organizational deeds on Earth. As you will see in this book, some of our most vital industrial resources, such as coal, iron, and limestone, owe their origins to chemosynthesis or photosynthesis followed by decay aided by microorganisms. The microbial activity involved helped pave the path to present life forms. Starch-laden grains we grow, such as wheat or corn, are energized by sunlight falling on their leaves and by nutrients from soil. Leguminous plants may be grown to enrich soil with nitrogen. Bacteria (called *Rhizobium*) on their roots remove nitrogen from the air and convert it into nitrates, which can be utilized by plants. Assisted by further microbial activity (from yeasts), grains may be converted into bread, sometimes referred to as the staff of life, or into beer. From such products we grow as self-propelled, sun-powered organisms, with built-in computers that we call brains.

The heart of the matter is this: The ecosphere is the creative source from which all life on our planet emerged. Since the sun is the vital source of heat energy and light energy upon which Earth life is dependent, it still fits the role of parent of the Earth and its teeming forms of life. In geological terms, this also fits the theory that the materials making up the planets of our solar system were once ejected from the sun, and are satellites held in orbit by gravitational forces.

While it may appear difficult to envisage oneself and ones neighbors of every imaginable species as bottles of sunshine, it is that energy within us that enables us to carry on functions from digestion to circulation, or from muscular activity to thought.

The Mood and Intent of "A Manifesto for Earth"

The May 2006 publication of *Proceedings* by the Canadian Association for the Club of Rome contained the text of a presentation given to members by Dr. Ted Mosquin. The title of his address was, "Some Thoughts on A Manifesto for Earth," and the following excerpt from his remarks indicate the perspective from which the Manifesto was written.

> As authors we saw ourselves as writing it from a distance…as if we were not participants in the processes of the Ecosphere…as if we were standing on the Moon while gazing at Earth and having in our

> minds the accumulated experience of two lifetimes of living on Earth plus the evidence of 200 years of science. When all was said and done, only 11 Principles emerged, no more, no fewer. These naturally and logically fell into two distinct categories; the first 6 are called CORE Principles. These describe the way things are on this Earth as witnessed by our experience, senses and findings, using the methods and instruments of science. The second group called ACTION Principles describe the consequential ethical obligations or duties that naturally flow from the first six. A guiding question repeatedly asked was "is a Principle missing?" In the end we found that we had six IS Principles and five OUGHT Principles. In philosophical parlance the Manifesto deals with three things. First, with ontology…or, what's reality? The Manifesto takes the view that our immediate reality is the Earth and its Ecosphere with all its organisms. This entire system is energized by sunlight. Second, epistemology…or, how do we know reality? The Manifesto says that it is by our perceptions that we understand the world…essentially through science and natural history. Third, it is about ethics: what then must we humans value most and how should we act on those values? The Manifesto is not based on contrived or 'other worldly' notions or assumptions about the reality in which we live. It is a secular document, as indeed it should be. Otherwise it could not be objective.

As I see it, "A Manifesto for Earth" offers humanity positive concepts and steps that must be taken if we are to make peace with Planet Earth. We should know by now that we have far-exceeded our rights and are in a precarious position. Political leaders offer only reluctant and inadequate compromises shackled to perpetuation of the potentially lethal status quo. No solutions have been offered that are as healing and hopeful as would be produced by implementation of the Manifesto we are commencing to explore.

Woodland Thoughts

My own life journey has led me to thoughts eminently compatible with the Manifesto's principles. My insights and sympathies have been derived from rural rather than urban places. From the sidelines where I live, I am sometimes reminded of a book written by Thornton Wilder. He wrote of a stampede caused by a single person shouting, "Run," and then acting on his own suggestion. Soon hundreds were running and shouting the same call of alarm: "Run." That, somehow, seems to be the current philosophy of modern confusion. Today it is no doubt an anomaly, but I have simply never felt it worthwhile to own a television. Quietness and reflection, which I enjoy, seem to be part of a more natural way of life, whereas television is all jangle and noise mixed with meaningless advertising, which appears as a travesty.

The cabin where I hang my hat is surrounded by forests. I am writing here of things that came to mind one winter evening when I was sitting at a table, reading by the light of a coal oil lamp. For several hours that day I had shoveled snow. After a simple sort of supper I returned to a book I was reading, the subject of which could be described as botany with a philosophical flavor. A window nearby was open a bit to neutralize fumes from the lamp and Gypsy, my German shepherd, was sleeping nearby.

Some of what I was reading had to do with the miracle of life and with photosynthesis, not a bad topic for a day that had been cold and overcast. Thinking about what I was reading, I became aware that there are probably fewer coincidences in life and more miracles than we commonly realize.

It was a slightly spooky night. Gypsy, moved closer to me, put her head on my leg and occasionally trembled. I put one hand on her head, assuming that she had probably got cold that afternoon. Inspired by what I had been reading, by the atmosphere, the stillness, and the dog, I pondered about my own understanding of life. What were the beginnings of the whole shebang and how did we come to be here?

I began thinking of the slowly rotating, incandescent sun, constantly ejecting photons from its whirling mass — light energy produced for billions of years. The freed photons raced away at the incredible speed of 186,278 miles per second. Just about incomprehensible! A small percent of the photons started on the eight-minute journey to the Earth, only 93 million miles away. (No, I wasn't thinking metrically, but in my native mathematical language). Two to four hundred million years ago, I reflected, in the period of time known as the Carboniferous years (including

Mississippian and Pennsylvanian periods and others), there were fern trees forty, fifty or sixty feet high, (fifteen to twenty meters high if you wish, but I wasn't there to be sure). Like modern ferns, the trees had enormous leaves, which contained the phenomenal substance called chlorophyll. Innumerable photons arriving on Earth crashed into chlorophyll molecules in the fern leaves and rent them asunder. Disarranged, the chlorophyll molecules rearranged themselves as carbon, hydrogen, and oxygen molecules, which were metamorphosed into glucose molecules. Glucose is the life-sustaining sugar that accounts for the growth in trees, the hawk stooping on the mouse, the leaping trout, the soaring swan and the stalking wolverine — energy even for the neurotic pilot of an automobile being driven madly to a rendezvous that never materializes.

Time passed. The huge fern leaves fell to Earth, decomposed, and returned to the elemental wealth of the world. They were buried by other organic matter, lost their hydrogen and oxygen through decomposition, and formed masses of carbon. Buried deeper and deeper and compressed as time passed, the carbon contributed to the fossil fuels buried out of harm's way in the Earth. These became energy sources to be used by not entirely provident people in what became an orgy featuring the insatiable consumption of energy. No sense of moderation safeguarded human enterprise.

I realized that when I took down and lighted my coal-oil lamp, and began reading by lamplight I had unearthed a conundrum. For my lamplight was actually stored leaf light, and before that it was emitted sunlight. And, wonder of wonders, even before that it was the energy of the Big Bang or of "Let there be light." Imagine that whether one sits on a granite boulder, roams along a mountainside, eats supper, or just daydreams, an individual is part of a greatly extended immortal moment.

The following morning I realized it even more, for Gypsy's trembling was explained when I went outside. Tracks in the snow showed that a cougar had walked around the cabin twice and stood beneath the window by the table. So there had been another form of energy only a few feet away. An amazing world!

Think about it. Aren't we some sort of miracles ourselves, mysteriously invited to the banquet of life? Aren't we perhaps supposed to rise above triviality? Should we be totally mundane and kneel at the feet of an economic convulsion that has seized the consciousness of mechanically addicted people who want to conquer the Earth? To conquer ourselves is the necessity, and to cherish the masterpiece of life and the infinite universe responsible for the phenomenon of our living, nurturing Earth —

that is the reality we must strive to achieve. The shepherds 'neath the stars who realized that our sufficient task would be to dress and keep the Earth may have been the wisest of wise men.

It is appropriate to understand something about our dependence on other organisms. We cannot make our own food from sunshine. Instead we must rely on green plants that are capable of producing their own food by photosynthesis, the world-supporting process energized by sunlight. The human economy is underwritten by the fact that photosynthesis, powered by the energy of the sun and catalyzed by chlorophyll in plants, produces some 300 billion tons of sugar or precursors of sugar annually. The process takes place on land and in the seas. The raw material produced in this manner is further utilized in other important syntheses such as the formation of amino acids from which proteins are built. Our own muscles, blood, bones, and nervous systems are nourished by food grown on the planet. Growth, bodily warmth, and the energy we use for whatever physical activity we undertake are made possible by plant energy made useable by the small fraction of solar energy Earth receives. Considering our relationship to other creatures, the Hudson River naturalist John Burroughs commented, "We are walking trees and floating plants."[2]

Though we may believe we are vastly superior to plants and physiologically absolutely distinct from them, hemoglobin in human blood and chlorophyll in plants are strikingly similar in composition. This is still another miracle, and it might be suspected that only a small (but enormously important) mutation may have been involved. And if you choose divinity as the catalyst, that is not surprising. Once life's enigmas are entertained, our technology is crude by comparison.

A molecule of hemoglobin contains 136 atoms of carbon, hydrogen, oxygen, and nitrogen in an intricate arrangement forming a central ring around a single atom of iron. In a molecule of chlorophyll, the same numbers of atoms form the same arrangement around a single atom of magnesium. Biologically we are not as distinct from plants as we think. Botanist Donald Culross Peattie observed that we could put a hand on the flank of a beech tree and reflect that we and the tree are relatives.[3]

Soil and Health

We also have a close relationship to soil. It bears reflection that we have, at times, been described as soil organisms. Not inappropriately, soil historian Edward Hyams, author of *Soil and Civilization*, subtitled the second portion of his book, "Man as a Parasite on Soil."[4] In addition to commonly known minerals such as nitrogen, potassium, phosphorus, and calcium, which are important to agricultural crops, soil may also contain minute amounts of "trace minerals" required for successful growth of specific organisms with special needs. Certain trace minerals are essential to plant and animal health. For instance, tiny amounts of iodine (an element) are needed for proper functioning of the thyroid gland.

Regarding soil and health, Louis Bromfield studied US Selective Service records from World War Two, and found that nearly 75 percent of young draftees from a southern state with badly leached soils were rejected as physically unfit, whereas Colorado, a state with undepleted soils, produced young men whose acceptance rate was the reverse of the state with poor soil. Just as our bodies need minute amounts of trace elements, so we can be sickened by too much of certain elements. Bromfield, for instance, pointed out that too much of the element selenium in soil vegetation has long been known to cause "alkali disease" in cattle.[5]

As a comment on Hyam's descriptive observation that we humans are parasites on soil, there is an interesting anomaly in the distinctive title, "parasites." Webster's dictionary defines a parasite as an individual who feeds himself at the table of another, and repays his host with flattery. In respect to this commentary, humans are inferior parasites, for while feeding ourselves luxuriously at Earth's bounteous table, we blatantly ignore our absolute dependence on the planet. Many people consider it strictly as a utilitarian resource and with great ignorance express a lamentable intent to conquer the Earth. In the human manner of *reductio ad absurdum* we are trained as consumers to "consume" the Earth rabidly without appreciation or concern for the health and well being of our generous host. We ravage the forests for fiber, much of which provides raw material for newspapers that stoutly present the homocentric, corporate view of the world. Corporations mine Earth for minerals to foster our economic aspirations and in the process disperse such minerals widely throughout the planet, eventually poisoning the soil that provides our food. Communities load oceans with human sewage and chemical refuse, thereby sickening oceanic ecosystems. Humanity's ten-thousand-year-old experiment in our mode-of-living at the expense of Nature, culminating in economic globalization, is already an absolute failure. The primary reason is that we have, with blind homocentrism, elevated the importance of our species above all else. A courageous change in attitudes and activities is urgent.

Our atmosphere has absorbed so much automotive and industrial flatulence that it can no longer qualify to be considered normal fresh air. Each year enough carbon is pumped into air to form a mountain a mile high and twelve miles in circumference.[6] *World Watch Magazine* states that carbon "is being unlocked and flung into the atmosphere at a rate of 7 billion metric tons" annually. The CO_2 gas that produces this amount of carbon is more like 27 billion tonnes — carbon being only part of CO_2. The weight of this 7 billion tonnes of carbon roughly equals the weight of a concrete block as tall as the Empire State Building and extending over 450 hectares.[7] Humankind has truly produced a surrogate or substitute world in which our host is succumbing to our monstrous demands. It is a biological axiom that "a good parasite is one that does not destroy its host." Yet that is precisely what our species is doing. Our whole stance as superior and unusually intelligent beings is strikingly egotistic. Scientists tell us that 99.99 percent of the species that have lived are extinct. Unless we change radically we will simply join them.

Few individuals realize that they live at the bottom of an ocean of air, which, though broadly referred to as the atmosphere, is also part of the ecosphere. Just as a fish hauled out of the water gasps and expires, so people would gasp and expire if hauled out of their supportive atmosphere. Everyone's complete reliance on the ecosphere is reason enough to recognize the priority that should be assigned to keeping it healthy.

It is now widely recognized that an altered worldview is imperative if people choose to become realistic about their tenure as a species. The key to this worldview is presented in the first core principle of the manifesto. It involves the change from a narrow, selfish outlook on life, to a broader, compassionate and survival-centered outlook on our membership in the amazing phenomenon known as life.

Three Worldviews

There are three worldviews that move progressively from absolute selfishness to a far more realistic selflessness.

The present worldview we embrace is described as homocentrism, which means that we are wholly focused on the well being of people and, as it turns out, particularly on the well being of important people. This species-centered attitude has enabled factory farming of animals for food. Poultry, cattle, and other animals experience human barbarity and ferocity as they are overcrowded, fed unnatural and revolting foods, and slaughtered for the all-important "market" in which the end goal is profit. The perversion of money has become the ubiquitous lust of humankind as it continues to weaken its grasp on survival. Our economic specialization is the unbearable straw on the human frame. The narrow view of homocentrism is comparable to looking through a slit in a wall, which permits a view of only those material things that bring immediate gratification. We have reached the apex of hedonism. Unfortunately education has its main concern in this area because society has narrowed its own goals to the materialistic and mechanistic, and has accepted the view that anything else (spiritual, aesthetic, compassionate, reflective) is counter-productive.

A second possible worldview is one that would be biocentric and extend empathy to other living creatures. By including other animals in humanity's concern, people would help themselves develop a more mature ethical stance. But this worldview is still limited because it ignores the Earth context from which all life has emerged. The flaw in biocentrism is that it will still encourage chauvinistic homocentrism to dominate thought, since we consider ourselves to be unrealistically wise and superior to all other animals. It is about time that we admit our nourishment is provided by Nature's generosity. Our importance diminishes each day as we dangerously disrupt Nature's order.

The more humble worldview, and the one most likely to sustain life, is ecocentrism, which recognizes and admits that the source of all lives on the planet is Earth itself, as empowered by the sun. This does not diminish religious views of life, at worst only clarifying their metaphoric qualities. The Earth, admitted to be life giving and sustaining, is clearly recognized in ecocentrism as productive of life in millions of forms, one of which is human. I have heard the thought, attributed to John Muir, that ecology is pulling up a dandelion and finding that everything else is attached.

Certainly a great mystery exists at the root of the universe. This could be more overwhelming in its meaning and plan than we will ever understand. There is truth to the ancient observation that we "see as through a glass darkly." Men have always pondered this enigma. For example, Empedocles, centuries ago reflected that, "God is a circle whose center is everywhere and its circumference is nowhere."

Many people seem absolutely fearful that there could be anything more important than their tiny selves. This inflamed ego may be a fatal flaw. In a survival sense, it would be true blessing for us to freely recognize that the matrix is more important than the species. In order to continue living, people must basically become wise, moral, and just and must also rise above self-centeredness to freely accept the idea of inseparability from other life forms. Like other life forms, humans have been preconditioned by their genetic inheritance. We occupy the "people" niche in a world where each species is qualified for its own role in life. There is the squirrel niche, the chipmunk niche, the owl niche, and as many other ways of life as there are other species. Organisms combine intelligence and intuition, abilities to communicate in different ways, distinct means of locomotion, inherited means of camouflage, and often unique characteristics especially fitting them for the roles they play in the fabric of life. Each organism is thus adapted to an ecological niche, which is a particular pattern of behavior and adaptation. The world of Nature and the roles of its occupants could easily be a life-long study. On the other hand, deliberate avoidance of the ongoing and continually amazing miracle of life is tantamount to casting a vote for species suicide. Humans are, first of all, of the Earth, and only uncomfortably and unwisely part of a much-ballyhooed sophistication that is hollow at its core. Is this pretense the actual root of the stress of life?

The Vital Importance of Other Species

"A Manifesto for Earth" speaks for all species when it states, "We humans are conscious expressions of the Ecosphere's generative forces, our individual 'aliveness' experienced as inseparable from sun-warmed air, water, land, and the food that other organisms produce. Like all other vital beings born from Earth, we have been 'tuned' through long evolution to its resonances, its rhythmic cycles, its seasons." From the foregoing thought let's consider another species, the earthworm, familiar to all, which has significant importance of which we are mostly unaware.

Learning about the earthworm and its importance to the world is an excellent way of "grounding" ourselves regarding the importance of both the soil and the earthworms so valuable in making and sustaining it.

Darwin spoke of earthworms as the greatest plowmen, more valuable than the horse, relatively more powerful than an African elephant, and more important than a cow in the services they perform for Earth. French scientist, Andre Voison in *Soil, Grass and Cancer* referred to the common earthworm found in Europe and North America as not only essential to good agriculture, but as the very foundation of all civilization.[8]

Earthworms are an important agent in the formation of humus, the brown or black material produced by decomposition of plant and animal matter, the organic part of the soil. In learning about humus one might seek the history of the word itself. The word humility is derived from the Latin *humilis*, which means "on the ground." It is not unusual to refer to a dependable individual as a person who has her feet on the ground. From this same root we get the words human, and humus. In the case of earthworms, their digestive tracts incubate thousands of microorganisms, which appear in the earthworm castings and become the activated base for humus. A USDA scientist estimates that on good soil, earthworms may produce more than five tons of castings per acre, per year. As well as assisting in the formation of humus, worms loosen soil as they move through it, mix and sift it, break up clods, carry down leaves, and provide avenues for plant roots. In early Greek times, Aristotle called earthworms "the guts of the soil." Either Voison's book referred to above, or Jerry Minnich's *The Earthworm Book* would be a valuable resource for further study.

Now why did I include the material about earthworms? The reason is that the little is big and the big is little. Life is wholeness. This is a difficult idea to grasp, but in oneness (the integrity) all the component parts have distinct functions. Earthworms are important to fertility, fertility is important to production of good crops, and good crops contain the nutrients important to health. Health can be undermined by missing essentials. For example, earthworm castings (their waste) are extremely valuable fertilizer. Heedless of Nature's own ways, our state of denial enables us to wipe out earthworms with the chemicals we use. We vote for the simplest ways to make profits and, step by step, move toward annihilation.

The Game of Life

Is ecocentrism more than merely a scientific technique for recognizing the importance of clay, sand, silt and gravel? Is it more than realizing the importance of rock dust, stardust, and every particle we encounter as we study life? Yes, because it is also part of a philosophy of wholeness. Ecocentrism recognizes that as much as we know, we do not know enough. It would advocate making haste slowly, which is one of the most famous of Latin axioms: *Festina lente.* Many may ponder why some people are in such haste to explore outer space when they know so little about inner space. We might ask why we are in such haste to fly or drive hither and thither when, for the most part, most people know quite little about their immediate surroundings. The ecosphere needs conscientiousness from our somewhat intelligent but highly erratic species.

In my mind, one of the great and expressive thinkers of the world was Thomas Henry Huxley (1825–1895). In an essay entitled "A Liberal Education," he made a comparison between learning chess and learning the game of life. He stressed the idea that if our life depended on our skill, we would very carefully learn the rules and appropriate moves in the game. Huxley described life in this way:

> It is a game which has been played for untold ages, every man and woman of us being one of the two players in a game of his or her own. The chessboard is the world, the pieces are the phenomena of the universe, and the rules of the game are what we call the laws of Nature. The player on the other side is hidden from us. We know that his play is always fair, just, and patient. But also we know, to our cost, that he never overlooks a mistake, or makes the smallest allowance for ignorance. To the man who plays well, the highest stakes are paid, with that sort of overflowing generosity with which the strong shows delight in strength. And one who plays ill is checkmated — without haste, but without remorse.[9]

The game we are playing today is not only for our own life, but also for the survival of the entire ecosphere — the biggest game yet!

[1] Winwood Reade, *The Martyrdom of Man* (London: Watts & Co., 1932) p.328.
[2] John Burroughs, *John Burroughs' America* (New York: The Devin–Adair Company, 1952) p.274.
[3] Donald Culross Peattie, *Flowering Earth* (New York: the Viking Press, 1939) pp.22–38.
[4] Edward Hyams, *Soil and Civilization* (London: Harper Colophon Edition, 1976).
[5] Louis Bromfield, *Malabar Farm* (New York: Ballantine Books, Inc., 1970).
[6] James Lovelock, *The Revenge of Gaia* (London: Penguin Books, 2006) p.73.
[7] Todd Neff, "Portraits in Carbon", *Worldwatch* (Jan./Feb.2007) p.19.
[8] Peter Tompkins & Christopher Bird, *Secrets of the Soil* (Earth Pulse Press: Anchorage, Alaska, 1998) p.41.
[9] Inglis, Stauffer, & Larsen, Editors, *Adventures in English Literature* (Canada: W.J. Gage and Company Limited, 1952) p.465.

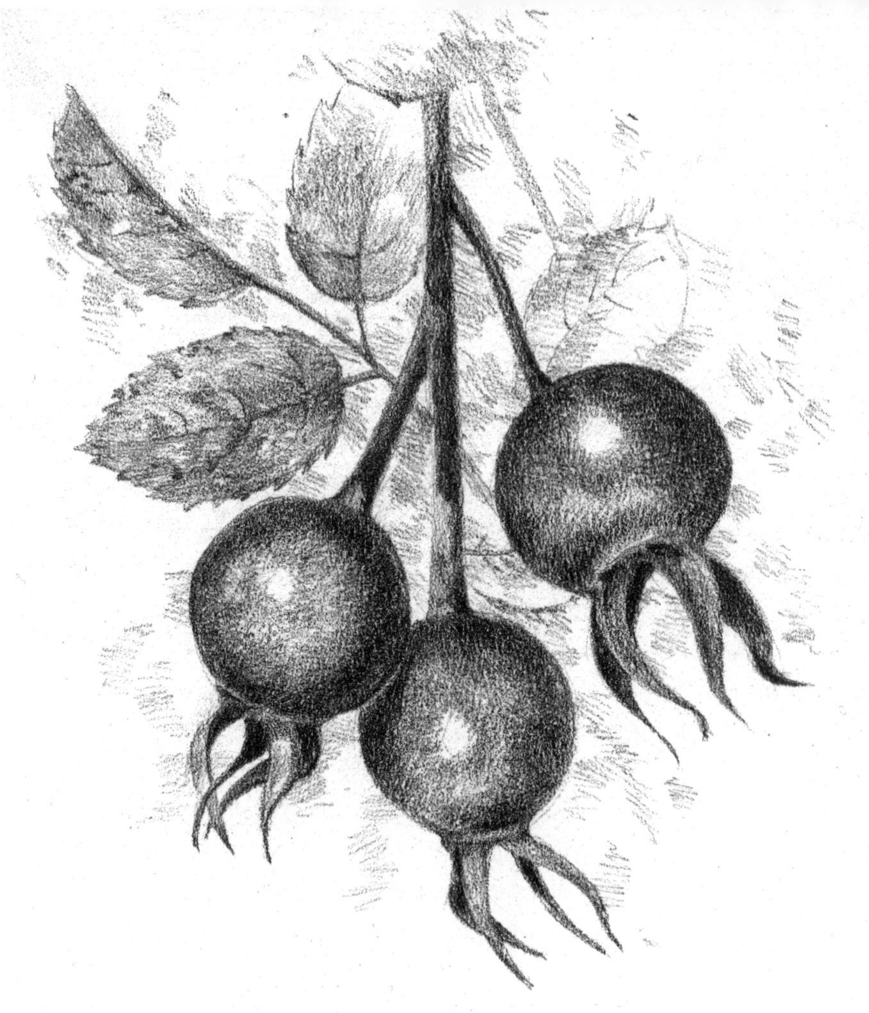

2: The Second Core Principle
The Creativity and Productivity of Earth's Ecosystems Depend on their Integrity

The word integrity is derived from the Latin integer, which means a whole number in mathematics. Integrity refers to the quality of wholeness, or of being in one piece. Earth's integrity has evolved over billions of years. It is a functioning, dynamic system because of the integration it displays.

But evidence is mounting that we lack sufficient wisdom to tinker with the established order of the planet without doing irreparable harm. Ancients knew that dualism, or separation, was the one sin. Gandhi accepted the Jainist belief that a person who conquers the sin of separation conquers all other sins at the same time. It is evident that our detachment from and lack of loyalty for Earth are sins of separation. This separation from the Earth lies at the root of ecological problems that now bombard us. We have convinced ourselves that our destructive economy is the mark of our genius, but it totally violates Earth's own economy. As a result ecosystems are reduced to shambles in order to produce goods that in many instances are totally unnecessary. Ponder the concept we call progress and realize that we are substituting an economic fantasy for the health and integrity of the planet. As society slides more toward inequality, there occur further separations between women and men, between rich and poor, between the powerful and the powerless, between weak and strong, between those who would conquer the Earth and those who would live in harmony with it, between religions and nations, and between various political convictions. It has been known for a long while that a home, a nation, or a world divided against itself cannot stand.

Earth is ancient. When its age is compared to a twenty-four hour day, humans arrived about thirty-five seconds before midnight. The 500-year-old Industrial Revolution and its offspring, our modern economy, occupy fifteen to twenty milliseconds of geological history. In that period of time human population pressure has increased dramatically, and the world's ecosystems are threatened with collapse. With our addiction to fossil fuel use, we may constitute a terminal disease of Earth. Knowing that we will also be terminated if we destroy our host, we should be initiating means of reducing our impact on the planet. Visible now are stirrings of Earth's last ditch, immune system resistance. This will result in storms beyond imagining, along with earthquake and volcanic activity such as has occurred in the far past. Our inertia will cause us to be eliminated or reduced to a sparse, forage-gathering remnant of present population numbers. Face the likelihood that if the planet succumbs we will be gone in a lose-lose situation. If we are eliminated by the planet it will be a win situation for Earth and a lose situation for us. Clearly we must bend and change in order to create the only possible win-win situation. That may mean hitching our cars to a fence post, walking or bicycling instead of riding, waiting until we get our own wings before we do any more flying, and writing letters or phoning relatives and friends instead if going a thousand kilometers to visit them. In England, not much more than a century ago, people often lived out their lives within fifteen or twenty kilometers of where they were born. The world will get along quite well even if we stay at home and do not visit all the continents or nations on the planet. We cannot live without Earth's support, but it got along without us for geologic ages.

We impetuously call our rudimentary improvisations "high technology" and ignore visible proof that Nature recycles most of her products and needs no garbage dumps. We bask in our supposedly unsurpassable intelligence, and myopically ignore Nature's marvelous inventions. One of these is visible in the antics, evasive maneuvers, and abrupt stops in midair that are standard operating procedure for hummingbirds coming and going around a feeder, or putting on the air brakes to hover within sipping distance of a delectable flower. And what incredible design accounts for the Herculean feats of ants carrying burdens equivalent to an average man packing off a two-ton automobile on his back!

We can only ponder upon the high technology displayed by *Convolvus roscoffensis,* a small worm that frequents beaches in the Channel Islands. This worm has algae within its body, which produce all the food they both need. The relationship (symbiosis) between the two species, an animal and a plant, is mutualism, a cooperative venture toward sustenance. To carry on photosynthesis and thereby make food, the algae require exposure to sunlight. *Convulvus* inhabits sand that is exposed by ebb tides. It rises upward out of the sand when the tide has ebbed, and lies in the sunshine while the algae make sugars and starches. It sinks into the sand when the tide returns so that it will not be swept out to sea, and it performs this ritual behavior without a brain. *Convolvus* offers a single example of untold numbers of intricate beneficial relationships shared by organisms.

These organisms are unaware that they represent Nature's own ever-present displays of production processes uniquely in harmony with one another. Some people however decry this as mere instinct and not comparable with our lofty intelligence, which has us pursuing militaristic dreams in space from a damaged planet sorely in need of all our tender loving care and attention. As a species though, the human race offers proof enough that it is our intellect, and not instinct, that has caused such massive degradation of the planet. If we continue to ravage Earth and cause our own extinction, perhaps further experiments with intelligence by the planet may be abandoned in favor of instinct abetted only by the non-speculative perception displayed by many successful species.

What We Leave Behind

Many thoughtful individuals have pointed out a serious example of our lack of integrity: it is that we continue to support the funneling of much of the mineral fertility of the land into the seas. It is blatantly counterproductive to eject colossal amounts of sewage wastes into streams, lakes, and eventually the oceans. Along rivers, upstream cities foul their waters, while downstream cities add chlorine and other chemicals to destroy the bacteria carried with these wastes. Demographer Paul Ehrlich pointed out nearly forty years ago that the water people drink from the downstream end of a river may have already passed through seven or eight other people.[1] The excrement produced in the United States is approximately five hundred thousand tons a day, not counting livestock wastes. This is a largely untapped resource that should be composted, and would have great importance for soil fertility.

Across Canada there are vast disparities in the way municipalities take responsibility for their sewage wastes. The Sierra Legal – National Sewage Report Card III, 2004 assigns top grades of A+ to cities such as Edmonton and Calgary that provide 100 percent tertiary treatment and UV disinfection of all sewage, with Calgary providing fully treated, free sludge to farmland as far away as Lethbridge. Using machines called terragators the sludge is injected into soil, to avoid odor. Reports state that farmers are "standing in line" for the free service. Calgary sewage processing is considered a cost of city living, and it justifies being called "state of the art." It demonstrates that processing and returning our own wastes to soil, where it belongs, can recover a valuable resource.

Grades received by some other cities in Canada will illustrate the responsibility or irresponsibility of various urban areas. On the east coast, Halifax received a grade of D. More than 65 billion liters of raw sewage are discharged into the Atlantic each year. On the west coast, Vancouver also received a D. It discharges up to 22 billion liters of combined overflow each year.

Vancouver has adopted a policy of procrastination, in that upgrades to merely *secondary* treatment will not be completed until 2030. The Britsh Columbia government, in its homocentric detachment from the ecosphere, gave a negligent extension of over twenty-five years to Vancouver and Victoria to allow them to continue irremediable pollution of the ocean. If its own integrity was improved by ecocentric comprehension, the provincial government could have prohibited the wasting of billions on the 2010 Olympics and attendant pollution, and instead spent the money to restore sewage nutrients to land mined of its fertility by over cropping, and poisoned by artificial fertilizers and pesticides. What both the provincial government's and Vancouver's sophisticated interest in the Olympics, and their disdain for taking care of its own sewage, reveals is that barbarism — in this instance, the neglect of cleaning up after oneself — takes precedence over fulfilling one's responsibility toward the Earth. This inability to put first things first is another strong indication that a basic preparatory course should be instituted for those wishing to qualify as politicians. Such a course would include such things as ethics, ecology of the planet, and public responsibility.

Victoria, British Columbia, the seat of the provincial government, received an F grade from the Sierra Legal group. Its sewage dumping into the ocean involves only primary screening, no treatment. More than 34 billion liters of raw sewage is still discharged each year. One wag suggested that the quality of government in Britsh Columbia is impaired because its legislative brains wind up in the ocean. Another individual commented that Victoria used to have a slogan, "Follow the birds to Victoria." This slogan was discontinued because a word rhyming with birds was often suggested as an alternative.

Canada's megalopolis, Toronto, was awarded a B-. Though it has secondary sewage treatment, it still discharges 9.9 billion liters of untreated sewage and run-off per year. It does have the toughest sewer-use bylaw in Canada. An F grade was given to Montreal, which supplies primary treatment only and exhibits no discernible intent to change.[2]

The above data offers at least a suggestion that the money being spent on war in Afghanistan is needed at home to clean up conditions in this nation. Our first need should be to stop making war on our planet by dumping wastes in the ocean that are valuable fertilizer for the Earth. Asiatic countries such as Japan, China, and Korea used human waste for agriculture as long as forty centuries ago. Romans also realized centuries ago that human waste is the rightful property of the land. Paradoxically we pollute rivers and oceans while organic matter in soil decreases. I recall a newspaper article years ago stating that Saskatchewan soils had lost thirteen tons of organic matter per acre. It was suggested that trains carrying cattle to feedlots might bring return loads of manure back to farmland.

The restoration of composted sewage waste to soils in Canada would help to return them to a natural state. According to the Manifesto:

"Integrity depends on intricate food chains and energy flows, on non-eroded soils and the cycling of essential materials such as nitrogen, potassium, and phosphorus." The Manifesto reminds us that, "ecosystems in their undamaged state...(as much as we can know them) still provide the only known blueprint for sustainability in agriculture, forestry, and fisheries." One thing that is known is that lands in North America, when first cultivated, were rich in organic matter. The return of organic matter to soil is the natural means of restoring fertility. Synthetic fertilizers are unnatural and may have served as a stopgap to cater to corporate profit, but they diminish the natural interplay of myriad soil microorganisms and destroy soil integrity. "The natural compositions of air, sediments, and water have been integral to Nature's healthy processes and functions. Pollution of these three, along with exploitive extraction of inorganic and organic constituents, weakens ecosystem integrity and the norms of the Ecosphere, the fount of evolving Life."[3]

The Cost of Our Wealth

It is clear today, though unwise, that mankind wishes to eliminate natural processes in agriculture, fisheries, and forestry for substitute methods that favor industrialization. Basically we seek to impose our will upon Nature and change the way it has operated for millennia.

Many books describe the teeming fish and wildlife that greeted immigrants to North America. Bison and antelope numbered in millions and supported healthy populations of predators. Fish in streams swam in such numbers that cast nets sometimes broke while hauling them ashore. Fifty-pound turkeys were shot from perches in trees, sometimes bursting their crops when they fell because they were so filled with food. Native Americans had not outstripped the prodigality of the land. The continent then displayed the dynamic health consistent with undisturbed natural integrity. Intricate co-dependencies and symbioses matured over millennia, and the living fabric of North America had become a verdant, fruitful, and seamless whole. Approached with respect and humility, the world designed by natural, ecological processes was one into which we might happily have fit if not confused by the illusion of our infallible intelligence and divine authority on this planet.

Subjected to our folly, fresh, clean air and pure potable water are now practically non-existent. Soils, likewise, have suffered faunal and floral simplification from pesticides and other pollutants poured onto them. The natural composition of air, water, and soils, once integral to Nature's healthy processes and functions, no longer exists on our planet. Future generations of all life forms will face the likelihood of tenuous footholds, due to rapidly altered life conditions caused by the haste and unreflective approach we have elected.

It is difficult to understand what we should do because we have been deliberately conditioned to keep out of the way of industry and let it do whatever it wishes. Governments have all too obviously left the fate of the land in the hands of big business enterprises, developers, miners, and recreationists who lease vast chunks of land in order to have a monopoly on recreational activities therein. In Canada wild land is referred to as Crown land, although irresponsible management suggests it should be called Clown land. Most rivers of any size have been dammed to produce more energy, although the truth of the matter is that we already have more energy than we can use wisely. If we abandoned new hasty technological improvisations to create a brief, lame sustainability and instead began a sustainable retreat from our incompetence, we might have a slim chance to survive in reduced numbers on Planet Earth.

Wisdom decrees that wherever there are substantially unimpacted ecosystems, we should adopt a hands-off policy toward them. This policy, combined with more realistic education, intelligent political control of corporations, population reduction, and governments focused on planetary health might slowly produce a wiser generation. Agriculture, forestry, and fisheries will always be the mainstays of survival on Earth; and the fourth primary industrial process the planet affords — mining — needs limitation, because heavy-metal toxicities and synergies offer substantial threats to life forms.

For all our verbose leadership, no government has ever had the sense to seriously study how we might exist on this planet in a harmonious manner. It is perilously obvious that we have overstepped our bounds and must work to restore the world we have dangerously altered. This would be infinitely wiser than continuing to destroy ecosystems worldwide. The principal war for years, as suggested above, has been war against the planet. Wars against nations are also wars against the planet because our fascination with weaponry enables us to blow up cities, schools, churches, and everything else we see. But, by the grace that peacefulness and deindustrialization might afford us, we might merit survival.

Double Ignorance

Our degradation of our planet is an example of what philosophers long ago entitled double ignorance. Ancient philosophers did not despair of the simple ignorance of "not knowing." However, when one knew better and still did a wrong thing it was called double ignorance. Some of the early impact on the planet may have taken place when we lacked knowledge that now exists. Our boast that we will conquer the Earth is abysmal double ignorance, verbalized. Our destruction of forests, oceans, and atmosphere was brought about largely because of our fantasy that we are favored children of the Almighty. The Earth, we assume, belongs to us, and our ego assigns us incomparable intelligence. We are just now beginning to realize that we will reap the chaos we have sown.

Unfortunately some of our more perverted corporations, Exxon and Philip Morris, for example, have been paying malleable scientists and non-scientists to publish material advising that problems we worry about, such as global warming, are non-existent. In Chapter 10 you will read an excerpt from a letter written by the Royal Society to Exxon about millions of dollars Exxon has spent providing the public with misinformation that denies global warming. This misinformation is given in spite of the decision made by 2,000 scientists of the IPCC (Intergovernmental Panel on Climate Change) that global warming is a real and serious problem. Such behavior by a major corporation brings into question the very existence of untrustworthy business giants so dedicated to their bottom line of profit that they will threaten all life on the planet. Corporate leaders should use what wisdom they possess to reverse their self-serving ways while they still have a chance.

On an even larger issue of unneeded pollution of atmosphere and soil, the world absolutely cannot afford the massive thermodynamic effect on climate caused by wars. It is time for us all to realize that what is left of the integrity of the planet is in extreme jeopardy. Peace might be a minimum price of human survival. A Quaker friend of mine speaks of the fact that it is a common belief of members of his faith that those who start wars should fight them. That would mean that President Bush and his cabinet should lead the troops in Iraq, just as Prime Minister Harper and his colleagues should lead the troops in Afghanistan. The troops themselves might be the corporate dragoons who make munitions and the petrochemical elite and others who reap the profits of war from their special watering places. If we could become a mature species and elect kindred politicians, warfare would be easily recognized as perpetuation of folly, both for people and planet.

The remainder of this chapter will give examples of ways in which peoples' unthinking actions have weakened and violated the natural integrity of the ecosphere.

(1) Algae are photosynthetic organisms (plants) that grow in oceans as well as on land. Their presence in oceans strongly influences Earth's climate. Algae extract carbon dioxide from air and use it for growth. They float in oceans and are sensitive to increased water temperatures. As global warming affects oceans, and water warms above ten to twelve degrees Celsius, their population decreases and the area suitable for them is reduced to cooler waters nearer the poles. Algae not only extract carbon dioxide but also produce dimethyl sulfate, a gas that oxidizes in air to form nuclei for cloud formation, which in turn enables seawater to remain cooler. Reduction of algal populations causes another of Earth's normal cooling mechanisms to be lost.[4]

(2) James Lovelock, an eminent scientist recently named as one of the world's 100 leading intellectuals, suggested that Earth-vegetation constitutes a living skin. Denuding Earth of forests and other natural ecosystems is comparable to burning the skin of a human. He contends that transpiration by forests and other vegetation is comparable to a human sweating. Few realize that mature trees on a sunny summer day transpire large quantities of water into the air. As water rises, it condenses to form clouds, which produce rain. Forests are guardians of streams and springs and are major stabilizers of global climate. Lovelock reminds us that when a human body loses 70 percent of its skin, death usually results. He commented in one of his books that the Earth has already lost 65 percent of its vegetative protection, and yet we continue our assault on it.[5]

Amusing Ourselves to Death

(3) Earth suffers from the voluntary dismissal of our own integrity and

the carryover effect this has on the vitality of all living beings. How has our integrity lapsed? As individuals with more understanding than we use, we let ourselves be superficial in the use of personal judgment. We readily accept the propaganda constantly projected by television, billboards, sales promotions, and credit cards that tempt us to buy now and pay later. Let's consider a classic example of temptation to which people easily succumbed. The following words were written by the sixteenth century poet Etienne de la Boetie. They were set down in reference to the far-seeing policies of Cyrus, the Persian king of the first century BC:

> When news was brought to him that the people of Sardis had rebelled, it would have been easy for him to reduce them by force; but being unwilling either to sack such a fine city or to maintain an army there to police it, he thought of an unusual expedient for reducing it. He established in it brothels, taverns, and public games, and issued the proclamation that the inhabitants were to enjoy them. He found this type of garrison so effective that he never again had to draw the sword against the Lydians.
>
> Truly it is a marvelous thing that they let themselves be caught so quickly at the slightest tickling of their fancy. Plays, farces, spectacles, gladiators, strange beasts, medals, pictures and other such opiates, these were for ancient peoples the bait toward slavery, the price of their liberty, the instruments of their tyranny.[6]

No doubt if Cyrus had waged war against their rebellion, the Lydians would have fought bravely and sacrificed their lives against the Persian armies, but they succumbed without a murmur to the subtle seductions that he engineered. Lacking self-restraint they became willing slaves, and devotees of hedonism.

The world of enticement that Cyrus created for the Lydians is prodigiously magnified today. Mass advertising keeps us informed of worldwide pleasures to suit every taste and to fit the dimensions of every pocketbook. People are tempted by alcoholic drinks and by drugs that will let them live in psychedelic euphoria. They are urged by advertising and tourist promotions to travel heedlessly in automobiles and aircraft to enrich industry, and to remain unfocused and on the move. Fashion is another trap for endless consumption. Urged to consume and live for appearances sake, people are encouraged to slavishly follow the latest styles. A patch in one's clothing is apparently more serious than a hole in one's soul.

People easily become addicted. In our search for pleasures we are not unlike addicts who must have another fix or die, even if they know that the next fix may be lethal. We know that people have survived because of Alcoholics Anonymous, and this depends on abstinence from alcohol. Today though, we need to cancel our addiction to unnecessary travel. "Automobiles and Aircraft Avoiders Anonymous" (and similar organizations) might help detach us from machines that are plagues on the ecosphere. It is a sad likelihood that humans have become so enslaved by their toys and entertainment that they will drive and fly steadily to their doom in preference to reducing their expectations. If we have the courage to put Earth health before our own idle pleasures we might go on to seek a workable manner of life on Earth. Unfortunately we lack both conscientious politicians and less obsessed industrialists. Cowboy lasso-twirler and humorist Will Rogers once told the public that Americans would constitute the first generation of people to drive to the poorhouse in an automobile. Now the prospects are far more ominous.

(4) British Columbia's role as host of the 2010 Winter Olympics may be looked at from the worldviews afforded by homocentrism or ecocentrism. From its inception the Olympic concept is pure homocentrism. Athletic abilities are tested in a ludicrously expensive and profligate manner. Thousands of athletes will be ferried through the skies, causing massive pollution, for a festival event that attracts many spectators who will themselves pollute their own way though skies and on highways to attend the event. These entire affairs costing billions and causing immense consumption of atmospheric-polluting fuels cannot be afforded by the ecosphere. Knowing now that global heating has reached the level of an emergency, Olympic spectaculars are truly counter-productive.

The Olympics, unilaterally opted for by British Columbia's Campbell government, constitutes a diversion of public funds from needed actions to the kind of spectacles that Cyrus the Great instigated. These events are meant to keep people amused and out of the way of politicians and corporations intent on running the world without concern for the wisdom of their actions.

Potential Energy Overhead

(5) Along with other massive violations of Earth's integrity by humans, one might look at mankind's self-serving efforts to remodel drainage patterns that have evolved over millennia. "Since 1950, the number of large dams, those over 15 meters high, has increased from 5,000 to 40,000."[7] There have been a disturbing number of dam failures causing substantial losses of life. These have resulted in immense damage to communities and to valuable farmland. Much dam construction has taken place for political reasons. The stated and unstated causes of dam construction include political prestige, profit, flood control, power generation, alteration of evolved drainage patterns, and, if the truth be known, employment for construction companies. Many dams have been built as a result of irrigation schemes, with costs paid by taxpayers who have provided free irrigation for land owned by multinational corporations. Bringing free irrigation water to California has made immense profits, often to benefit vested interests. Much of California would otherwise be a desert.

The book *Cadillac Desert* by Marc Reisner offers evidence of political intrigue, and even of knowingly building dams in unsafe locations. Terror propaganda spread thickly by politicians to keep the public fearful and docile, never reveals the enormous, ever-present, terror potential, biding its time in the immense quantity of potential energy stored in enormous, artificial lakes tentatively impounded behind dams. Many of these man-made colossi loom above too many cities, which they may ultimately wipe off the surface of the planet. Politicians noisily seeking to remove terror threats should review the extreme havoc that dam failures might cause. Hindsight will ultimately reveal that many dams should never have been built and need to be removed.

The 2,600 people who died when a mountainside avalanched behind Italy's 262-meter-high Vaiont Dam, remind us what can happen when unstable land along an impoundment avalanches. On October 9, 1963 a landmass of 240 million cubic meters slid into the reservoir behind the Vaiont dam displacing 50 percent of the water. The event took thirty seconds and the mass moved at the rate of twenty to thirty meters per second. Tsunamis, up to 300 meters high, traveled in both directions from the slide. The concrete dam survived but two villages to the west were destroyed, while to the east water swept over the dam burying the town of Longaroni beneath seventy meters of water.

Reisner speaks of the failure of Teton Dam in Idaho, a dam known to have had dangerous fissures in the abutment area before it was built. He states that a member of the Geologic Survey told a group of friends that the site of the Teton dam was "really a crummy spot to put a dam," and a geologist had commented that there was no true bedrock for the dam. Anyway, on June 3, 1976 the dam failed. "One second there was a dam, three hundred feet high and 1,700 feet wide at the base; the next minute it was gone." The flood from the dam, billions of tons of water, wiped out Wilford and Sugar City, Idaho. By the time the flood reached Rexburg "the real damage to Rexburg was done by Sugar City and Wilford. Reduced to giant pieces of flotsam — silos, walls, automobiles, telephone poles, pianos, trees — Wilford and Sugar City were a battering ram afloat, smashing Rexburg to pieces." What the waters did to farmland was disastrous. Tens of thousands of acres were stripped of soil down to bedrock. It was later noted that more land was destroyed and made incapable of ever growing crops again than would have been provided with irrigation by the dam. The integrity of the Earth is often ravaged by "development." It is also obvious that homemade terror is frequently caused by human ambition. "Four thousand homes were damaged or destroyed, fifty businesses were lost" at Rexburg. Only eleven lives were lost due to heroic efforts by state police. However, if dam failure had taken place at night rather than in daytime, the downstream communities would have been overwhelmed without warning.[8]

It is understood that the existence of various Earth structures, such as mountains, deep lakes, oceans, and valleys are compensated for by adjustments to isostacy, general equilibrium between topographical masses and the underlying supporting material.

Sometimes isostatic changes may be quite abrupt. For example, rapidly changing water levels in reservoirs often trigger readjustment tremors. Readjustment tremors as they are called are actually small earthquakes as rock strata respond to new pressures when a reservoir is periodically brought to full-supply level. Such tremors have caused failure of some dams, with resultant loss of life, and the sum total of all these new weight distributions no doubt stresses major faults and perhaps increases the magnitude of earthquakes in other locations. When Lake Mead behind Boulder Dam in Colorado reached its full depth of 475 feet, the weight of water was twenty-five billion tons. As a result, over 600 tremors, the largest being magnitude 5, were registered in ten years in an area that had not felt a shock for fifteen years.[9]

On December 10, 1967, a severe earthquake resulted from filling the reservoir behind 103-meter-high Koyna Dam in India. The reservoir was initially filled in 1964 with two billion cubic meters of water weighing two billion tons. A first tremor had been felt in 1962 when the reservoir contained 850 million cubic meters of water. A stronger shock was felt in 1965. In 1967 a scientific paper read at a congress on large dams claimed the quakes were caused by crustal readjustments that would decrease and eventually stop. The prediction was wrong. The December quake registered 6.4 on the Richter scale and caused damage throughout India; some 200 people died in the village of Koynanagar, southeast of Bombay. Prior to building Koyna Dam the entire Bombay–Koyna–Poona area had not been subjected to tremors.[10]

Quakes greater than magnitude 6 Richter and severe damage have been caused by reservoirs filling in Rhodesia and in Kremasta, Greece where 480 houses collapsed and 1,200 were damaged. One fatality occurred and sixty people were injured.[11]

The examples above illustrate an optimistic characteristic of human nature, which is to believe that man-made change is within the realm of our control. The recommendation to think positively often means putting one's head in the sand, in order not to think of the ugly possibilities that might occur if things go wrong. Dam failure releases instant chaos, and the cause may be a sharp, unexpected tremor in an area where tensions have been amassing slowly.

Both Mica Dam and Revelstoke Dam in British Columbia have

slides in unstable conditions upstream from the dams. Tom Thompson, an expert on earth-filled dams, advised a hearing in Revelstoke that if Little Chief Slide above Mica Dam should avalanche and put a wave over the top for thirty seconds, the dam would fail. In his words this would bring about "a catastrophe of continental proportions." Downie Slide, above Revelstoke Dam (a concrete dam) contains over a billion cubic yards of unstable material. Though some stabilizing work has been done, the slide continues to creep downward in response to the law of gravity. Once again note that "stored terror" is a part of "progress." A number of communities lie along the Columbia River below Revelstoke. After the Columbia River enters the US, it flows more than 700 miles before it reaches the Pacific Ocean, passing through numerous communities along the way. If unleashed by a dam collapse, freed water will head downhill, sweeping communities, highways, and rail lines along with it.

Part of the problem that underwrites the construction of all dams is the enthusiastic but often questionable human demand for more energy. Historians and scientists have stated emphatically that we have more energy at our disposal than we have the wisdom to manage. To speak of flood control by building a dam that might fail and cause a flood with a magnitude beyond imagination is at least a trifle ridiculous.

Governments have developed charades that are referred to as environmental hearings concerning mega-projects such as large dams. These are incorrectly called developments although they are in effect a direct rebuttal of centuries of natural development. The preferential attitude favoring often dangerous, so-called "development" allows serious objections and flaws in these projects to be glibly diffused, swept under the table, and considered trivial. As a result we arrive fifty or more years later at the realization that we have created threatening situations that are denied, resented, or excused as acts of God. As consummate actors, our power brokers call for "studies" to defer action, or pretend surprise when tragedies take place. A project manager of a large dam, when asked about the fate of people downstream if the dam should fail, answered his questioner with the remark, "There has always been a price on human safety." Hopefully there are now some signs that many individuals are awakening. Unfortunately, humans have been reckless and have delayed so long that the likelihood of major ecosystem failures is probable. In the final analysis our vaunted intelligence has revealed itself as little more than a potentially suicidal propensity for tinkering.

Dams violate the integrity of the Earth by allowing the evaporation of as much or more than a cubic meter of water from reservoir surfaces in sub-tropical or tropical areas. In arid or semi-arid regions where evaporation rates are high, water loss "is typically equal to 10 percent of the reservoir's storage capacity."[12] Dams also destroy valuable farmland due to flooding of fertile areas frequently found alongside lakes that are to be turned into reservoirs.

Building new dams also causes climate changes. One might note that the nearly 700-square-mile area flooded behind WAC Bennett Dam in British Columbia stored enough heat to produce milder temperatures than were common. To be sure, global warming is usually associated with fossil fuel combustion, but large bodies of stored water moderate local and regional temperatures. In the past, when very cold winters occurred, pine beetles were destroyed. Now the mountain pine beetle has destroyed lodgepole and jack pine trees on nearly nine million hectares of forestland in British Columbia. Milder winters have allowed them to thrive and to wipe out huge areas of forest. The forest industry, obtuse in its endless quest for more wood, is now clear-cutting not just pine but spruce and all other available trees, creating a huge desert with its forest and soil destroying machinery. The stodgily homocentric government has apparently never realized that really standing on guard for Canada would give the health of land first priority. Reminded of the statement that "Man proposes but God disposes," it might be noted that man proposes many schemes that lead to the destruction of ecological balances that are centuries old. Obviously building of the Bennett Dam helped to increase the effect of global warming.

There is a lot of truth in the concept that we should undo the havoc we have wrought with natural drainage systems. Not too far away from the Bennett Dam is the huge Nechako reservoir created to provide electricity for aluminum production by Alcan. The dam destroyed beautiful Ootsa Lake. If this dam failed it would wipe out Prince George, keep the Fraser River from flowing for a week or more in order for its channel to accommodate the Nechako River overflow, and when all that water reached the mouth of the Fraser River there would be some rather serious effects on Vancouver. Nechako Reservoir is another climate modifier.

Certainly a bevy of experts will contend that the dams mentioned are foolproof and will endure until the cows come home. But the cows may be on their way. There have been many more earthquakes in recent years, some of high magnitude. There have also been a number of smaller quakes stemming from the Madrid Fault on the Mississippi River. Don't overlook the Madrid Fault. In December 1816, the Madrid Fault in Missouri produced the strongest earthquake in US history. It was felt in an area of one million square miles and seriously disturbed 30,000 to 50,000 square miles of land.[13] The area was sparsely populated at the time. But the extent of the earthquake suggests that man's construction efforts may yet be subject to a higher authority. As Robert Ingersoll once pointed out, "Nature does not deal in rewards and punishments, but only in consequences." There are other strong Earth forces imminent, such as the restless caldera in Yellowstone Park, the awakening of volcanoes in the Coast Mountains, the recurring quakes along the San Andreas Fault, and plenty of others. A great make-work project in a renovated, ecocentric economy would entail removal of numerous, questionable dams and return of rivers to their natural rhythms. Yes, we could live with less energy.

There seems to be considerable merit in Robert Service's disturbing conclusion that we, "chop and change, and each fresh move is a fresh mistake."

Economist Lester Brown reminds us that when the Japanese attacked Pearl Harbor, President Roosevelt immediately ordered a ban on the sale of private cars, rationing of gasoline, fuel oil, sugar and some other substances and nationalized industries to produce goods needed to meet the wartime emergency. Brown states flatly that the ecological emergency today is greater than the one to which FDR responded.[14] We desperately need a new breed of politicians and a new economic outlook. Brown was clearly calling for a limit to fossil fuel use. This is a critical need and some politicians probably realize it. But our society's addiction to gas-guzzling vehicles is acute. I have spoken of the idea of rationing fuel in several talks that I have given. On each occasion some individual has commented on the fact that he would rather be dead than have to give up his driving, or that it was his personal right to use as much gasoline as he felt like using. This is indeed a serious and life-threatening addiction. Where is the politician who will give a national address and begin the effort to bring people to grips with the truth that we subsist by the grace of planetary integrity, which we are mightily threatening?

We are asked to give loyalty to many things, to families, communities, states or provinces, to nations and allegiances of nations, to one or another religion, to the economy, to political parties, to sports teams, to countless organizations; but never, it seems, to the Earth. Each of the loyalties we are asked to give demands our wholeness and dedication. Since this is impossible, I would say that our first loyalty should be to the planet, which is a manifestation of whatever divinity may exist. When people sing "Praise God from whom all blessings flow," they should realize that those blessings flow to our tables or homes through the abundance of the Earth. To have personal integrity we must scrutinize our deeds and intentions, and ask first how our actions will affect the integrity of the planet.

[1] Paul R. Ehrlich and Anne H. Ehrlich, *Population, Resources, Environment* (San Francisco: W.H. Freeman and Co., 1970) p.126.

[2] www.sierralegal.org/reports/sewage_report_card_III..pdf

[3] "A Manifesto for Earth," *Biodiversity,* Volume 5, No.1 (January–March 2004), p.5.

[4] James Lovelock, *The Revenge of Gaia* (London, Eng.: The Penguin Group, 2006) p.51 & 160.

[5] James Lovelock, *Healing Gaia, Practical Medicine for the Planet* (New York: Harmony Books, 1991) p.153–186.

[6] Quoted in *Manas,* Vol. 39, No.16, (April 16, 1986) p.1–2.

[7] Lester R. Brown, *Plan B* (New York: W.W. Norton & Co. Inc., 2003) p.33.

[8] Marc Reisner, *Cadillac Desert* (Vancouver/Toronto: Douglas & McIntyre, 1993) pp.404–408.

[9] Gordon Ratray Taylor, *The Doomsday Book* (Greenwich, Conn.: Fawcett Publications, 1971) p.37.

[10] ibid.p.35.

[11] ibid.p.36.

[12] Plan B, p.33.

[13] *New Age Encyclopedia* (Lexicon Publications, Inc. 1980) Volume 6, p.188.

[14] *Plan B,* p.203–206.

3. The Third Core Principle: The Earth-centered Worldview is Supported by Natural History

Natural history was one of the earliest subjects to appear in literature. Hippocrates, Pliny, and Aristotle were among those who studied and pondered upon Nature. Pliny's *Natural History,* though erroneous in places, is a classic. Over the years a vast assemblage of natural history studies has appeared in print. A main purpose of this chapter will be to provide synopses of a number and variety of comprehensive natural history selections, which offer introductory insight into a worldview that includes species other than our own.

As I look back upon my own education I realize that courses devoted to natural history would have offered me a better model for living more wisely and with far greater understanding of the diversity and dynamism of Earth. As a result, this chapter is focused on the appropriateness of natural history studies as an important part of education in schools and as adult education material. For people with a yen to move closer to Nature's ways, natural history writings are delightful.

To be sure, elementary students are exposed to some natural history. Their fascination with frogs, butterflies, and various plants demonstrates instinctive interest in other living beings. Their enthusiasm verifies the appropriateness of such study. Natural history, in fact, is the parent of ecology. Over the years thousands of nature tales have appeared in print and graced many classrooms. In 1868 a German biologist, Karl Reiter, named the branch of biology having to do with the relationship between organisms and their habitats *Oekologie*. The following year his countryman, Ernst H. Haeckel, popularized the word, which became known as "ecology" in English. This science, it has turned out, embraces virtually all other sciences and has enabled humans to grasp more fully the varying structures, habitats, and adaptations of thousands of creatures belonging to the phenomenon of life.

As human awareness of sciences increased, more need was established for specialization in specific aspects of particular disciplines. I once attended a scientific lecture in which the speaker punned that eventually we might see individuals studying to become experts of diseases of the left thumb or of the right ear. Nicholas Murray Butler, president of Columbia University, in a commencement address defined an expert as "one who knows more and more about less and less." A newspaper queried this by asking if the ultimate result of such expertise would be to know everything there is to know about nothing? Obviously the schoolchild accumulating general knowledge is an optimum candidate for the legions of natural history tales that are a basic part of historical thought. Knowledge at an early age forms a foundation for life and should certainly include awareness of the basic structures and forces at the root of life itself. Rachel Carson and John Burroughs are two authors who had a splendid ability to teach natural history.

At an academic level anthropologist Margaret Mead touched on the value of general knowledge in remarks she made when she was appointed president of the Scientists' Institute for Public Information (SIPI). She reminded her colleagues that there is a great need for generalists to compensate for the increasing number of specialists being turned out by universities. Obviously we now need generalists who can help restore wholeness. As citizens, voters, and political candidates, students with an ecology background fostered by natural history studies will have greater appreciation, rapport and understanding of life than schools now provide. A broader base for informed citizenship would lead to more wholesome goals than mere accumulation of material possessions.

Unfortunately modern education is homocentric and increasingly technocentric. It only superficially considers the long history of development that preceded and underwrote our emergence as a species. As mentioned, an understanding of natural history led to the development of ecology, which reveals much that is ignored about the mutual interdependence of all organisms that depend on Planet Earth for their continued well-being. Our modern technological single vision constitutes a materialistic vacuum which distains spirituality and threatens all life. Ecology, because of its earth-centered focus, is a natural taproot for developing holistic education in place of what is now structured mainly to benefit mankind's unrealistic political and economic expectations. In spite of all attempts to make us accept the role of children of technology, we are earthlings and can only survive by making Earth the center of our concern. By re-arranging the current school curricula wisely, nature study will become a part of school studies long before human history is taught.

Students would benefit from clearly understanding that the world is primarily a place to live, not merely a site for the transaction of business. The ancient goal of dressing and keeping the Earth and of planetary stewardship was not actually too trivial for us; we were, and still are, too trivial for it. If young people were taught to understand that we are a dependent part of life along with other organisms, they would develop a realistic, ecocentric comprehension of the world. This principle of the Manifesto recognizes that natural history studies in schools would constitute an extremely valuable resource for redevelopment of Earth consciousness.

Dr. Stan Rowe, a lifelong educator, in his book *Earth Alive* discusses the unfortunate likelihood that education today is an "outworn system" intent on fulfilling human desires "at the expense of extermination of other species, pollution of water, air, deforestation, the degradation of soils, and expanding deserts." He cites wrong values and beliefs as the crux of the problem. One of the chief erroneous assumptions is "the conviction that humans have the God-given right to dominate, control, and manage the entire Earth and all that is in it." It is notable that Christian religions are silent about such an attitude, although the Book of Leviticus, in the Old Testament, refers to God's statement, "for the land is mine; for ye are all strangers and sojourners with me." Leviticus (25:23). Considering the damage humans have wrought on the planet, religions have been too long silent and remarkably indifferent to what is taking place.

Dr. Rowe asks the question, "Should values be taught in school?" Answering his own question he states, "Inescapably they always are." The values we need, he contends, are ones that are in "harmony with ecological truths." In his words, "The duty of education is to foster a new frame of reference for all that people think and do, a reorientation toward deeper and truer insights with power to set this and succeeding generations on a more charitable and creative path vis-à-vis the surrounding world."

He goes on to say, "The historian and educator Hilda Neatby recounted a story about Lincoln Steffens, editor and author, who traveled to Europe just after his graduation from university in the late 1800s. The turning point in his education came when he met a group of Oxford men and listened to their discussions. 'Those men never mentioned themselves,' he said. 'Their interest was in the world outside of themselves…their conversations…established in me the realization that the world was more interesting than I was.' Neatby added, 'Here is as good a definition of education as any: the discovery that the world is more interesting than oneself. It is also a good definition of citizenship, and of mental health.'"

Summing up his recommendation for future education, Stan Rowe writes: "In revising school curricula, both perspectives, that of the planet-as-whole and of people as potential participants within it, cry out for primary attention. The two together, comprising human ecology, should be the core of education. Without the revolutionary understanding of human ecology, civilization will continue on its self-serving and suicidal way. Without the guidance of ecological comprehension all traditional curriculum subjects are likely to prove useless, posing dangers to sustainable living on the planet."[1]

The Web of Life — An Ancient Concept

Consult the *Meditations of Marcus Aurelius,* (ca. 150 AD) and you will read, "Always think of the universe as one living organism, with a single substance and a single soul; and observe how all things are submitted to the single perceptivity of this one whole, all are moved by a single impulse, and all play their part in the causation of every event that happens. Remark the intricacy of the skein, the complexity of the web."[2]

The foregoing quotation would be an excellent topic for analysis in a student essay, (also not a bad exam essay in a teacher education course), and is part of the evolution of human thought. Today such awareness is suppressed by propaganda from our economic leaders. Profit seeking has obliterated concern for the health and sustainability of the web of life, while ecological literacy is trivialized in politics and business.

In an organic evolution course I learned that many animals became extinct because of over-developed specializations. Fossils reveal that the prominent fangs of saber-toothed tigers sometimes grew so long they locked the jaw, and Irish elk with their massive antlers developed curvature of the spine and could barely hold their heads aloft. The professor wryly commented that people are candidates for extinction because our inflated egos exaggerate our intelligence and "we erroneously assume that no creatures other than ourselves are important." He referred to this attitude as an unconscious form of pre-extinction behavior.

Society still labors under Rene Descartes' insistence that only humans have souls, and all other living creatures are machines. It may come as a surprise to many people that this idea contradicts the Book of Job (12:8–10) in the *Old Testament,* which refers to "the soul of every living thing," and also advises, "speak to the Earth, and it shall teach thee."

It is also interesting to consider the stringent religious outlook held by the Jainist group of Buddhists. They attributed souls and life to not only animals and plants, but also to the four original elements: fire, air, earth and water — a basically ecocentric outlook. The founder of this sect, Vardhamana Mahavira, around 600 BC proposed what might be looked upon as a single commandment: "Do not injure, abuse, oppress, enslave, insult, torture, or kill any creature or living being."

Ecology — A Taproot for Education

Experience, education and reflection convince me that ecology is the realistic taproot for education. After all, our mental, physical, psychological and spiritual health is dependent on the health and integrity of the Earth. The word ecology, by definition, means "study of the household." Thus our homes are part of the planetary household. Education should foster *thorough* awareness of our relationship to Earth and should also cultivate responsible, caring behavior toward the wholeness that sustains us.

Understanding that we are part of the integrity of Earth will help dispel the unfortunate illusion that Earth is our property and can be expected to adjust its billions-of-years-old natural laws to suit people's unreal economic expectations. Our insistence on an increasing Gross National Product disallows any reflection about the Gross National Disaster visible in collapsing fisheries and climate, vanishing forests, and overstressed ecosystems. A constant barrage of pro-machine propaganda reinforces the hypnosis of machines. This has led to disdain for labor-intensive, less injurious ways of doing things. Thus we suffer helpless dependence on technological devices at a time when energy costs are becoming prohibitive. This technological dependence was noted by behavioral zoologist, Dr. Konrad Lorenz, former director of the Max Planck Institute, who coined the term "mechanomorphism" to describe increasing human addiction to machines.

By awakening students to their natural heritage, they may influence their parents. Inasmuch as teacher education basically ignores Nature as the starting place for life, teachers could enhance their own education by selecting readings that broaden knowledge of the world that surrounds us. By analogy, rooting themselves in Nature through study would result appropriately in starting their work from the ground upward. Only by knowing more about the complexity of the world that sustains life can they teach others to keep their feet on the ground. It is not foolish to speak of life today as frivolous — as disconnected from roots. Considering the state of the world, it is imperative to provide material encouraging people of all ages to adopt an Earth-centered worldview. We may make a great fuss about space launches and the sensationalism they provide, but our destinies will be worked out on Earth. At this time particularly, our leaders have no time for grandstanding and stupefying the public.

The hour-long CBC radio interview with British scientist Dr. James Lovelock on July 13, 2006 conveyed his conviction that we may have only thirty years left before catastrophic population reduction occurs; at most, a century. While other scientists do not agree with Dr. Lovelock's timetable, his recommendation that young people need to be educated to live in a world where basic needs must be met largely by one's own efforts, does make sense.

Ministries of education and universities should awaken to the need for courses such as "Natural History for Teachers," which will serve to increase ecological comprehension.

If I were structuring such a course, the book *Microcosmos* by microbiologist Dr. Lynn Margulis and artist Dorion Sagan would be useful as a background text. Subtitled "Four Billion Years of Microbial Evolution," it would help to fill a vital knowledge gap in the education of most adults, and enable them to recognize their interdependence with multitudes of other organisms.

Some of the facts *Microcosmos* reveals are rather staggering, to wit: "More than 99.99 percent of all the species that have existed are extinct, but the planetary patina, with its army of cells, has continued for more than three billion years." You might note from that sentence that the normal chance of a species surviving is 1 in 10,000; and we aren't doing too well. Too many vital facts we need to know have been omitted from curricula.

Another passage in *Microcosmos* explains, "Our own bodies are composed of ten quadrillion animal cells and another *one hundred quadrillion* (100,000,000,000,000,000) bacterial cells." Not only does the microcosm continue to evolve around and within us; the authors remind us, "the microcosm is evolving *as* us."[3]

These excerpts indicate the value that *Microcosmos* would have as background for any reader. The book offers important historical and biological perspective. As you learn that bacteria "invented" fermentation, the wheel (the proton rotary motor), sulfur breathing, photosynthesis, and nitrogen fixation, you will realize that immensely higher technology than we now have, preceded us by millions of years.

The deeper one penetrates natural history literature, the more one becomes aware of life as kindred, interrelated organisms enjoying their common moment in the sun. And as an individual progresses from simple tales of the lives of individual species to more profound studies, the concept that we are members of a miraculous web of life becomes more and more an unshakeable truth. We will have our hands full to become good members of the Earth community of life into which we were born.

On a Piece of Chalk

Here is a thumbnail sketch of a natural history essa
ers and students insight into the immense importan
ly insignificant thing as a piece of chalk, someth
classrooms. The essay also illustrates how nicely natural history readings may be integrated with well-written literature and current events. The essay, entitled "On a Piece of Chalk" was given as a lecture by Thomas Henry Huxley to the Working Men's College in Norwich, England in 1868.[4]

He told the audience that they might dig under their feet and find a layer of chalk many hundreds of feet deep. This layer could be traced for 280 miles, and on the shores of Kent it constitutes the white cliffs of Dover.

Huxley explained that microscopic study of a cubic inch of chalk shows thousands of skeletons of sea-dwelling microorganisms, the most common of which is named Globigerina. These are predatory marine animals that look like a blob of jelly with tiny tentacle-like fibers. Globigerina have existed for hundreds of millions of years. Their skeletons (called "tests") are made from carbonate of soda. When Globigerina die, their skeletons rain down to the ocean floor forming ooze, which eventually becomes compressed into chalk, a form of unconsolidated limestone. Studies reveal that Globigerina cover forty-eight million square miles of ocean bottom. Chalk produced by them may be found almost worldwide. It underlies Central Europe, and is present in the Crimea, in Syria, and in the mountain ranges of Lebanon. Globigerina ooze covers the bottom of the Atlantic Ocean and forms a great underwater plain stretching from Valentia, on the west coast of Ireland, to Trinity Bay in Newfoundland. Chalk deposits, made up of sea-dwelling Globigerina may also be found high up on the flanks of mountain ranges. To geologists this is unquestionable evidence that seas once covered lands uplifted throughout the long history of the Earth.

Guided by their teacher, students reading "On a Piece of Chalk" will learn that oceans have long provided partial solutions to excess carbon produced on Earth. Earth stores carbon on land in vegetation, notably in trees, and in the oceans as the shells or skeletons of thousands of marine species. Forests, marine plants and animals, chalk and limestone are part of Nature's management of excess carbon and its effect on our climate. Much carbon produced on Earth by volcanoes, fires, and by respiration

living organisms has been absorbed by oceans. It has been impossi-
to know exactly how much carbon oceans absorb each year.

Huxley's essay offers great insight into the history of the Earth, adds to geographical awareness, and stimulates thinking about our own origin and the dramatic events that have slowly changed the planet over millennia. This accumulated knowledge, if studied in our youth, would give us the wisdom to avoid the blunders that have been created by modern technology.

More than a Marketplace

We have a truly great need to see ourselves as part of the life on Earth, and to accept our common inheritance with other life forms. Poet laureate Alexander Pope realized this long ago, stating in his *Essay on Man,* "All are but parts of one stupendous whole,/Whose body Nature is and God the soul." (Epistle I, Line 267)

Secondary school sciences, succumbing to economic and political pressure, have focused so heavily on preparing students for disciplines such as medicine or engineering, that each individual's personal membership in the universe has been ignored. This personal relationship has been left as an irrelevancy in a society intent on profit and development. The world has been reduced from a place wherein love, joy, and beauty might reign, to nothing more than a marketplace. This is a contrived reduction to absurdity. Our lives very likely depend on regaining pride in our membership in life and our myriad relationships with other living things. The world is not a lonely place at all.

It is an arguable hypothesis that in the generations since World War I adults, and particularly children, have lost many aspects of their birthright. Unfortunately, young people have not been consulted and have not willingly sold their birthright, as Esau sold his to Jacob for a mess of pottage. The birthright of which I speak is a complex thing. As Earth beings, children should enjoy fresh air to breathe, clean water to drink, healthy soil from which their food can be produced, and adults who will seriously and diligently protect these natural treasures bequeathed to all living things by the universe. For many generations these qualities of their being were safe for all living things.

Once, the majority of children grew up in villages or other small communities where they could safely run, play, and absorb Nature as their primary setting. For many, there were swimming holes and fishponds nearby. There was bird song and sunshine, a garden for fresh food, animals for pets, and in general many natural pleasures, which nostalgia will recall to some people. It is a different story today.

Genetically programmed for adult care in a safe, secure, and natural setting, half the world's human offspring are coping with urban environments where stresses and hazards can be extreme. They may enjoy vicarious experiences on television and have many toys, but even those pleasures that adults feel are safe may be less than that. One of the hazards of "progress" is that products are vigorously ballyhooed and put into use long before their safety has really been assessed. For example the June 2006 issue of *Harper's Magazine,* in a regular feature concerning scientific advances, notes, "Experts warned that childhood exposure to television and video games should be seen as a major public-health issue and warned that these media should come with a health warning, like cigarettes."

Since natural history is the story of Earth and its diversity, the scientists who produced "A Manifesto for Earth" advise that, "Stories of Earth's unfolding over the eons trace our co-evolution with myriad companion organisms through compliance, and not solely through competitiveness."

Although we seek to escape identification with "animals," it is interesting that the Latin root of the word animal is "anima" which is defined as "soul." The idea of the immortality of the soul was a favorite theme of Pythagoras. The Roman poet, Ovid, reiterated the idea that the soul roams to and fro, adopting whatever form it chooses from beast to man and never dies. Birth to Pythagoras was the undertaking of a different, new beginning and death merely the occurrence of change. He despaired of human carnality and envisioned a golden age long ago when abundant fruit and food from harvests enabled people to live "without staining their lips with blood." He lamented that men of his day could find it in their hearts to kill their "plough-mates" (oxen) and eat their flesh. He urged, "Abstain! Never by slaughter dispossess Souls that are kin, and nourish blood with blood."[5] If one met with a predatory animal that sought his own life, it was justifiable to kill the animal in self-defense, but the predator should be left without being used as food.

Since Nature has been around a lot longer than our species, schools would be following the established order of Earth history if students received instruction from at least a representative selection of writings on natural history. These date from the time when people were just beginning to keep records and continue to the present.

The Study of Life

Internationally acclaimed natural history writer Joseph Wood Krutch noted that modern laboratory sciences have remained severely detached from wholeness. Focused on dissection and anatomical study, they have lost awareness of their subjects as living, feeling beings. The attitude of detachment makes it difficult for some professional individuals to admit that such unique sentience exists. Now, at a time when fascinating studies are revealing our very close kinship with both animals and plants through common heredity, there has been a decided reluctance to accept such relationships. Would it be unbearable for us to lose our sense of superiority?

A good first natural history essay for students to study as an introduction would be one entitled "Basic Forms of Life" by Joseph Wood Krutch. It appears in his book, *The Great Chain of Life.* The essay begins by discussing the ancient concept of "spontaneous generation." This was the belief that the simpler animals found in mud and slime and in rotting organisms simply began life from the corruption of such substances. Krutch quotes Samson's observation that, "Out of strength cometh forth sweetness."[6] Samson had seen the carcass of a lion that apparently rotted itself into a colony of bees. Tracing history, Krutch reports that spontaneous generation was widely supported even through the eighteenth century.

The same essay will teach students how to make a hay infusion and, with a microscope, see dried life reappear from the wisp of hay inserted in water. For a first essay in a natural history course, this one will offer vivid evidence of life reviving itself in the truly tenacious fashion through which it has become present nearly everywhere on Earth. He explains that if the wisp of hay is boiled and sterilized first, no such life will appear. A reader will learn that the simplest animals digest without stomachs, take in oxygen without lungs, secrete uric acid without kidneys, and collect it in pockets wherefrom it is excreted, without bladders. Furthermore they exchange heredities without sex organs.

In a book entitled *The Best Nature Writing of Joseph Wood Krutch* you can read an essay entitled "The Mouse that Never Drinks." This is an account of the kangaroo rat, which is much more mouse-like than rat-like.

The kangaroo rat is an example of an animal adapted to live in deserts where water is absent. The largest kangaroo rats live in Death Valley, California where the length of time between rainfalls may be eighteen months. They have a special ability to metabolize water from the seeds that make up their diet. Their urine is twice as concentrated as less specialized animals and their feces contain 45 percent water rather than the 65 percent that is normal in other rodents. I talked with a scientist who had studied these animals and he had dug up burrows and found that during the heat of the day the nocturnally active animals rested at depths where soil temperatures were sixty-eight to seventy degrees Fahrenheit. In this way they avoided surface temperatures sometimes over 100 degrees Fahrenheit during the heat of day. He found that due to temperature lags their burrows were cooler during the day than they were at night. These animals propel themselves by jumping on their hind legs, hence the word kangaroo as part of their names. Also, they proceed in hops as much as a foot (thirty centimeters) high, and as they jump they switch their long tails thereby making an erratic course, which must distract would-be predators.[7] There are dozens of essays by this author and many of them offer valuable insights into plant and animal behavior. Krutch is a classicist in the quality of his work. Many libraries have his books.

The Father of Medicine

Hippocrates, still referred to as the Father of Medicine, was born in the fifth century BC. An essay entitled "On Airs, Waters and Places" indicates that, in his time, serious study of natural conditions was recorded. Hippocrates had many ecological insights. He advised, "when one comes into a city in which he is a stranger. He ought to consider its situation, how it lies as to the winds, and the rising of the sun; for its influence is not the same whether it lies to the north or the south, to the rising or to the setting sun." He also thought it was important to consider the water source of the city, whether the water was soft or hard, whether it came from high, rocky places or whether it came from a marsh. Hippocrates showed that respect of and appreciation for Nature existed long ago, and likely exceeded the acuteness of many modern individuals. His remarkable early awareness of ecological conditions is one reason why his essay would be a valuable addition to the training of teachers as part of a course in natural history especially designed for them. In the sense that natural history includes the unfolding of our own development, this piece is an historical landmark. One place in which it may be located is in volume 10 of the 54-volume set of *Great Books of the Western World* (1952, Encyclopedia Britannica, Inc). The same essay read to secondary school students would fit as an interdisciplinary combination of science and history. It is remarkably understandable and its wealth of concepts is instructive and interesting.

Human Impact on Earth

A background in natural history will help individuals to be aware that human history rarely focuses on human impact on Earth. A modernization of education would take care to interweave and integrate the many factors that collectively create the successes and failures of individual cultures.

The definition by Edward Gibbon that "History is indeed little more than the register of the crimes, follies, and misfortunes of mankind," certainly reveals an undeniable aspect of history. But it also suggests that the average man, unlike Hippocrates, does not care to understand the ease with which humans can degrade regions of the ecosystems in which they live. Natural history will broaden the base from which we can determine our relationship to other living things and to the physical world. Our efforts to conquer Nature clearly express ignorance of our dependence on it. Our realization, from study, experience and reflection, will show us that we are parts of a wholeness, and develop the mature realization that cooperation with the natural world would be more fruitful than the "me first" attitude the human species has adopted.

Many clear examples of the crimes and follies imposed on the health and vigor of natural ecosystems are provided in the book *Topsoil and Civilization,* written by Vernon Gill Carter and Tom Dale.[8] This book can be considered a classic and provides excellent examples of human impact on Earth. Let us look for a bit at what Carter and Dale can tell us of the history of ecosystems in Mesopotamia.

Today we have accelerated forest destruction beyond any pace previously known in history. To learn from history we might look at the consequences of ruthless deforestation in one of the world's famous garden spots. This was in the area that is historically referred to as the Fertile Crescent, once a treasure land of marvelously fertile, silty soils formed by the Tigris and Euphrates rivers in what was then known as Mesopotamia. The fertile valley lands combined with the water of the Tigris and

Euphrates rivers created a form of natural wealth that made Mesopotamia a cradle of civilization. Part of this area, now known as Iraq, once made a major contribution to early civilization and agriculture. However, early in history the Armenian Hills were deforested. The destruction of forests caused great quantities of soil to be washed from the hills. This silt choked the irrigation ditches, which were used to water the excellent but dry soil that was being farmed. Studies have shown that vast amounts of silt accumulate in rivers following removal of vegetation from the hillsides. For instance, in New Mexico, the Rio Puerco showed a proportion of forty-two cubic inches of sediment to fifty-one cubic inches of water. (Statistics from ancient history in Mesopotamia are unavailable).

The fertile soil in Mesopotamia led naturally to repeated invasions. Barbarism was then, and remains, a constant threat to stable civilizations. At various times the Sumerians, Assyrians, Babylonians, Seleucids and others controlled the land. Slave labor was used to clean the irrigation ditches. In 637 AD and for more than a century thereafter, Mesopotamia was a province in the Moslem Empire. Following more unrest, the Mongols eventually reconquered the area in the thirteenth century.

The Mongols disdained agriculture, which had been carried on for five millennia. They hated farmers and cities, and set about to demolish both. They destroyed the major Nahrwan canal and a diversion dam that fed it, which led to loss of control of floodwater of the Tigris. Water then periodically flooded out of control, dissected the land, and eroded much of the fertile soil. In the twentieth century, following more floods and razing of forests, the land fed less than a quarter of the number of people it fed in ancient times.

The sort of history indicated above is not typical of history taught in schools. Humans acting upon Earth still bear testimony to the words of the French diplomat and literary celebrity Francois Rene de Chateaubriand, "Forests precede civilizations and deserts follow them."

People, today, mainly inhabit cities and are far removed from the sights, sounds, and surroundings afforded by rural life. Education, to be comprehensive and to fill a growing intellectual void, can find only in natural history, a sufficient body of knowledge to focus on the real world, and thereby to ameliorate the effects of a mechanical world that has set its goal in materialism. It is now evident that the world has been crippled by the onslaught of technology that is not guided by wisdom. In fact, the pace of events, inventions, and lack of foresight that exists should be an incentive to make youth of the present aware, more concerned, and more ethical than the generations that have preceded them. We can no longer afford to smugly sacrifice Nature's long established and complexly interrelated order for a simplified process of extraction and consumption in order to accommodate our profit-dazzled economy. Nature has been forcing upon us a reality that has been caused by the unreality of our assumptions. Although we frequently utter a cliché that "the individual or nation that ignores history is doomed to repeat it," we rarely let such realistic wisdom be our guide. We are lacking in knowledge of the intimate processes that established the world beneath our feet. Completely immersed in our substitute world of round-the-clock entertainment and promoting our own dreams, we cannot recognize ourselves as one more of Nature's species playing out what appears to be a short-term role on the planet.

Learning About the Real World

Natural history is the story of the Earth itself, and is a very broad and extensive account. It has been constructed from observations, studies, compilations, and reflections made by vast multitudes of individuals, over lengthy periods of time. It is, of course, a partial history since the history of the Earth dates from times long before human beings existed, and there may well come a time when our species has departed, and some other, perhaps wiser, species has replaced ours. Some of the best educational fare that students in schools might digest would fit under the banner of the history of the Earth.

In an urbanized society, the provision of natural history readings and discussions in classrooms would be both highly educational and refreshing. The expression of connectedness to the world we inhabit would be one means of reducing the anomie of individuals. It seems at odds with current thought to call the endless pursuit of money, fame, and power a religion but it has all the dedication of many religions and a fanaticism peculiarly its own. This view will ultimately collapse and leave many individuals wounded, particularly if their hearts are in their pocketbooks. Much of today's desperate need for mental therapy could be better solved with a hoe in hand than by a psychiatric appointment.

Education today is tailored to fit the mayhem of a world in which citizens are endlessly exhorted to fill the role of consumers. The wealth of thought and study that has produced a large body of material in natural history suggests that it is a far more significant form of history than any other. The study of natural history would enable a “minds on” form of reading and reflection offering significant improvement of vocabulary and literary skill, along with the realization that we are full-fledged members of the family of living brings.

Another essay, which would incorporate biological and behavioral information about our fellow beings on Earth, is one entitled “Driver Ants.” The essay was written by Thomas Belt (1832–1878) and is an excerpt from his book *The Naturalist in Nicaragua,* written in 1874.[9] It would be informative to precede reading of the article by study of a globe to learn the location of that country; and it would be useful to peruse an encyclopedia to gain some general knowledge about Nicaragua. The article contains a behavioral study about a species of foraging, hunting ants, the Ecitons, which are abundant throughout Central America. The author compares the behavior of these hunting ants to the behavior of races of humans who were once hunters. Both human and ant hunters have to move from one location to another in order to survive.

Not only does the author describe the hunting behavior of the ant, he also describes the behavior of other insects, some of which are able to save their own lives by standing perfectly still. Spiders sometimes spin a single thread and save themselves by dangling from it. He describes simple experiments exemplifying ways in which the ants communicate with one another, and the social behavior by which they rescue one another from harm. He also observes cooperative behavior through which ants form bridges with their own bodies. The essay offers convincing evidence of intelligent behavior on the part of these ants, and notes that birds accompany the ants to feed on other insects flushed by the ant hunters. This article is one that can easily create new awareness of the larger world so often taken for granted.

One of the numerous, inadequately explained events of ancient history is the emergence of mammoths that have thawed from glaciers. An excellent description of the discovery of a mammoth is found in “Mammoths in the Flesh,” an excerpt from the book *The Mammoth, and Mammoth-Hunting in Northeast Siberia* (1926) by Bassett Digby.[10] As a reporter and author, Digby traveled extensively in Siberia and witnessed the thawing of these animals from their icy tombs.

The selection includes historical evidence of the discovery of these huge animals. Digby asks us to imagine an elephant-like animal about thirteen feet high and fifteen feet long, with tusks eight feet long and legs about one-and-a-half feet thick. The tale he recounts offers interesting evidence of the edibility of mammoth flesh, which was eaten by wolves and bears, and also speaks of the superstitions concerning the giant mammoths that were held by Samoyedes, Ostiaks, and other tribes who lived in the land where these frozen mammals were found.

Among the natural history essays that offer superb material for educational insight is one called "The Seeds of Life and The Sleep of the Seed." This essay is taken from a book entitled *Flowering Earth* by Donald Culross Peattie. The essay provides a basic and fascinating account both of the history of life on Earth and of the amazing structures and processes that account for the abundance of plant life. The reader will also learn about autotrophs, early plants that fed on iron and other compounds. Part of the fascinating history of the Earth is that before there were green plants, there were self-feeders (autotrophs), which lived by oxidizing iron. Life itself, Peattie points out, is "one vast oxidation, one breathing and burning." One of these self-feeders, known as *Leptothrix,* "literally eats iron."[11]

Let's digress for a moment. The process by which *Leptothrix* feeds itself is called chemosynthesis. Modern green plants feed themselves and the animal life they support by photosynthesis. When *Leptothrix* thrived, it lived in a darkened world of water. It was so effective in making iron its table fare, that it formed the vast Mesabi Range, the most important iron range in the world. The Mesabi has been so important that the Sault canal, through which its iron is shipped to steel mills in Pennsylvania, transmits a greater tonnage than any other canal in the world.

The foregoing is just a bit of the very pertinent knowledge provided by Peattie's discussion of the seeds of life. Unless you are a well-educated botanist, you can learn very much from this essay. Teachers will also be able to tell their students what to look for to find *Leptothrix.* These bacteria are still with us, a primeval form of bacteria that poses absolutely no threat to life. Drivers of our many cars probably owe thanks to *Leptothrix* for providing the steel for the manufacture of automobiles.

Oh yes, there will be new words to stumble over, but there will be new awareness of topics we all should know about. Life is learning and learning makes one more competent to understand the world. Natural history is the real history of the world and certainly of what makes life possible. Reading, spelling, vocabulary, and science, the origin of life, the transmittal of life, and all sorts of respectable educational objectives would be met in studying this article. Ambitious students might well choose to read the entire book from which this selection is made.

The Extermination of the American Bison, by William T. Hornaday, was published in 1887 as a report from the US National Museum. It is a valuable historical document in that it reveals the horrendous impact that modern man, with his weapons and his propensity for slaughter, can have, even on a species that numbered in the millions. At that time the Indian depended on the buffalo for food, clothing, shelter, bedding, saddles, ropes, shields, and ornamentation. One record I have read talks of "sportsmen" on railroad flatcars in the United States shooting the animals for sport. One group shot and killed twenty-seven bison from a flatcar, tarried long enough to savor a meal of bison flesh, and left their slaughter. Bison were shot for hides alone, were massacred as a means of depriving native tribes of a food supply on which they were dependent, and were killed merely because herds sometimes blocked railway lines and impeded the flow of traffic. By starving native tribes, the natives were forced onto reservations where beef and blankets substituted for the freedom they had enjoyed. Hornaday pointed out that the government could have restricted the killing of bison for far less money that it was spending "to feed and clothe those 54,758 Indians."

Hornaday included information on the impact of killing bison in the USA and the effect it had on the tribes in Canada. He quoted Professor John Macoun and his book, *Manitoba and the Great Northwest* (p.342): "During the last three years (prior to 1883) the great herds have been kept south of our boundary, and as the result of this, our Indians have been on the verge of starvation. Whereas the hills were covered with countless thousands (of buffalo) in 1877, the Blackfeet were dying of starvation in 1879…During the winter of 1886–87, destitution and actual starvation prevailed to an alarming extent among certain tribes of Indians in the Northwest Territory who once lived bountifully on the buffalo. A terrible tale of suffering in the Athabasca and Peace River country has recently (1888) come to the minister of the interior of the Canadian government…In the form of a petition…It sets forth that 'owing to the destruction of game, the Indians, both last winter and last summer, have been in a state of starvation. They are now in a complete state of destitution, and are utterly unable to provide themselves with clothing, shelter, ammunition, or food for the coming winter.'…Heart-rending stories of suffering and cannibalism continue to come in from what was once the buffalo plains."

Obviously the preceding natural history material has much to say about the conduct of our species, about its attitude to the life of other animals, about morally reprehensible ecological behavior, and about random, reckless, and equally reprehensible behavior toward indigenous people and the assemblage of animals that provided them with sustenance. The information also shows that native tribes originally lived in harmony with the bison; that their impact on bison did not interfere with the bison's prodigious population, which would likely have been sustainable for centuries.

As a natural history reading of abiding value, *The Sense of Wonder* by Rachel Carson helps to create just what its title suggests. This book describes Miss Carson's relationship with her nephew Roger. I would say that it also describes what should be every young person's introduction to the world. It is the sort of book that a family should read together and could also use to initiate some of its own activities. The author says, "If I had influence with the good fairy who is supposed to preside over the christening of all children I should ask that her gift to each child in the world be a sense of wonder so indestructible that it would last throughout life." Her own sense of wonder is infectious, and she contends that such a sense of wonder serves to eliminate boredom and reduce the disenchantments of later life. With such an attitude a child would reject the "sterile preoccupation" with senseless acquisition of possessions that now dominates the world. Above all, the young person would not be alienated from the natural world, which provides endless stimuli to enhance our strength.[12]

An extremely thought-provoking and readable account of our relationship to soil may be found in the writings of John Burroughs (1837–1921). Known as the Hudson River Naturalist, Burrough's twenty-five books examine many facets of the natural world. Burroughs was both a naturalist and a philosopher, and it may be seen that the study of natural history provokes the sort of pondering and reflection that leads to philosophizing. Unfortunately the pace of modern life stems from an attitude of utilitarianism that permits little time for thought.

A worthwhile reading from *John Burroughs' America* is a chapter entitled "The Soil in Ferment." For young people or for anyone who enjoys literature, the author's writing is graphic and charming. When he writes that, "We are rooted to the air through our lungs and to the soil through our stomachs," he makes vivid the concept that we are utterly dependent on the air for our breath, and likewise dependent on the soil to provide the food we eat. People often neglect to realize that the meat we consume is itself produced from the soil-grown grass that feeds the cattle, or other animals, whose lives are sacrificed to our food preferences. Soil also, as Burroughs points out, is the grist (grain) from which our bread is made. If for no other reason, we need to read about seeds and soil because we cannot live without eating, and every bit of our food is dependent on the functioning of the fundamental creations of Nature.

There is a beauty and a sense of reverence imparted by "The Soil in Ferment." There is also an unspoken reminder that we have departed dangerously from the wisdom to recognize that our individual lots and houses are but nests or nooks in which we live on our earthly home. Thoreau, years ago, put life in a more subtle perspective, when he pointed out that our porches are, in effect, the entrances to our burrows. This may be not only an interesting homonym to the name John Burroughs, but also a reminder that even though our houses are above ground and are fabricated, they still perform the function of the simple burrow that provides shelter for our families.[13]

Livestock judge, professor, and Lieutenant Governor of Alberta, Grant MacEwan is widely known for his writing. In *The Best of Grant MacEwan,* the essay entitled "Loaned for a Season" points out the need for change in the relationship between people and the Earth they occupy: "More than any other animal, man is a stranger and foreigner in Nature's community. He's the one guilty of leaving parts of the Earth in waste; he's the troublemaker, the one inspiring the most fear and terror in fellow creatures. If the wild things could speak, they'd agree; the sight or smell of man is enough to send any wild animal to flight." Specifically, he explores the meaning of the word conservation and reminds us that no matter how rich is a land, the things we casually call resources are not inexhaustible and have more value than we assume. He also advises us that we have a responsibility to those who follow us, and that we should not live in such a way as to deprive our heirs. Leaving things better than we found them amounts to close adherence to the Golden Rule.[14]

A Sand County Almanac, written by Aldo Leopold, is an environmental classic. The half-century that has passed since it was published points with increasing urgency to the need for wide reading and discussion of the essay "A Land Ethic." In fact the entire book is worth reading and discussing.[15] Readers will find few books as crowded with insight. Leopold, himself, pondered the need for revised education. In spite of conservation oratory, which existed in his lifetime, he noted that it was mostly "letterhead piety" and insincere chatter. In other words, it was mostly talk without corresponding action to care for the Earth. As he expressed it, for every good step taken publicly, there were a couple steps taken backward on "the lower forty," where people couldn't see what was happening. Leopold suggested that the root of the problem had much to do with something that was missing in education. His reaction to the idea of increasing the volume of education was that it made more sense to alter the actual *content* of education. This suggests a truth we have been evading — the now very visible need to change the content of education, to start teaching people about their absolute dependency on the Earth. No mere cosmetic effort will suffice. Actually this is a need in education that has never been addressed. It is obviously absent in the background of politicians and the business leaders who drive the economy by obliterating the productivity of the real world to make their substitute world "profitable." The word "profit" may well be the most misleading word in our vocabularies. For example, "profit" is the motive behind the cutting down of our forests. Yet by doing so we are actually made poorer because these same forests are responsible for stabilizing and protecting the climate, guarding runoff so that streams will flow year round, preventing soil erosion, and a host of other measures which make the world livable. Profit is reaped at the cost of excessive damage to Earth.

A Closer Look at Soil

Living Earth, by Peter Farb, was published in 1959 but remains an excellent introduction into teeming and complex life found in soil. Readers will learn, perhaps for the first time, about springtails and mites, two of the most common organisms in forest soils, and their presence by the thousands in each square foot of soil. They will learn about shrews, which Farb calls the "fiercest animal on the globe," and about the ability and willingness of these mammals to take on and triumph over animals many times their size. They will learn about the force called imbibition, which enables a seedling to push its way through an asphalt road. I remember seeing mountain ash trees one to two meters high that were growing on an abandoned stretch of highway. The book offers valuable learning on almost every page, and is another educational resource for people of all ages.[16]

Secrets of the Soil, by Peter Tompkins and Christopher Bird, includes a chapter called "Microcosmos," which is about the "farmer's unpaid workers" that range from microbes to clearly visible organisms. The reader will learn about organisms that collect nitrogen from air, ones that make important chemical changes, others that are scavengers, and even ones that parasitize other microbes. To the thoughtful individual it will raise the question of immense damage we may do by adding pesticides and other pollutants to the soil.

Another chapter in the same book is called "Biomass Can Do It."[17] This chapter raises the interesting possibility that composted human sewage can return wasteland to productivity and achieve the same purpose with marginal lands that are infertile. It entertains the promising possibility that by living with Nature and cooperating with it we might be able to raise adequate quantities of food to sustain society and reduce fuel use simultaneously. In America, as much as 300 gallons of oil per acre is consumed by agriculture that is dependent on machines, artificial fertilizers, and pesticides. Ninety percent of all grains grown in the US wind up as livestock feed designed to sate cultivated appetites for animal protein. Unnecessary waste results from 25,000 calories of energy (plant food) being needed to grow 1,000 calories of beef protein. A good portion of the meat we eat putrefies in the human gut. By reducing meat intake, human health would be improved and fossil fuel use decreased.

The foregoing selections offer convincing evidence that very fascinating material is available to provide better education for young and old, and thereby return us to the realization that the miracle of life far outstrips the ravaging of Earth for profit. A new idea of wealth emerges, which is simply that the health and integrity of the planet is the real wealth of life, and that by living in cooperation with the innate order of the Earth we would be head and shoulders above the world of competition.

Actually, natural history and ecology provide a wealth of knowledge, which corroborates the wisdom of life within Nature's bounds that was established before our Johnny-come-lately species adopted the view that we are God's gift to the world. We would not be able to exist if it were not for the hundred quadrillion, or thereabouts, bacterial cells that inhabit our bodies, along with ten quadrillion animal cells that shape our being. As microbiologist Lynn Margulis states in *Microcosmos* (page 67), "The life forms that recycle the substances of our bodies are primarily bacterial." We are being kept ignorant of the truth that the life exterminating chemicals poured into the world by our industries are weakening and sickening the most fundamental structures of our bodies — for the sake of profit. When governments, to please industry, establish "tolerances" of so many parts per million (ppm), or parts per billion (ppb), they imply that tiny quantities of pesticides and poisons are harmless. This is not true. Their proclamations to this effect are political, and politics is characterized by one departure from truth after another.

Basically an Earth-centered worldview is the natural outcome of study in which a basic knowledge of natural history provides innate awareness of membership in the family of life. Life is supported by Earth, and returns to Earth upon death. Being aware of the amazing miracle of being, and knowing the source of that being, removes much of the sense of meaninglessness that threatens many people wary of the trivial pursuits of our times. To realize that we are cousins to all other living things is no little accomplishment. It is also an understanding that power structures choose to deny.

~~~

[1] Stan Rowe, *Earth Alive* (Edmonton: Newest Press, 2006) p.103–105.
[2] Marcus Aurelius, *Meditations* (Great Britain: Penguin Books, 1975).
[3] Lynn Margulis, *Microcosmos* (New York: Summit Books, 1986) p.15–49.
[4] William Beebe, *The Book of Naturalists* (New York: Alfred A. Knopf, 1944) pp.131–150.
[5] Ovid, *Metamorphoses* (Oxford and New York, Oxford University Press, 1998) pp. 356–360.
[6] Joseph Wood Krutch, *The Great Chain of Life* (New York: Pyramid Publications Inc. 1966) pp.13–27.
[7] Joseph Wood Krutch, *The Best Nature Writing of Joseph Wood Krutch* (Richmond Hill, Ont.: Simon & Schuster of Canada, Ltd., 1971) p.60–71.
[8] Vernon Gill and Tom Dale, *Topsoil & Civilization* (Norman, Oklahoma: University of Oklahoma Press, 1974) pp.39–55.
[9] Beebe, *The Book of Naturalists,* pp.151–159.
[10] Ibid, pp.335–340.
[11] Donald Culross Peattie, *Flowering Earth* (New York: Viking, 1939) pp. 205–217.
[12] Rachel Carson, *The Sense of Wonder* (New York: Harper & Row, 1965).
[13] John Burrough, *John Burroughs' America* (New York: The Devin-Adair Company, 1952) pp.270–287.
[14] *The Best of Grant MacEwan,* edited by R.H. Macdonald (Saskatoon, Sask.: Western Producer Prairie Books, 1982) p.58–62.
[15] Aldo Leopold, *A Sand County Almanac* (New York: Oxford University Press, 1966) pp.237–263.
[16] Peter Farb, *Living Earth* (New York: Harper & Row, 1959) pp.1–30.
[17] Peter Tompkins & Christopher Bird, *Secrets of the Soil* (Alaska: Earth Pulse Press, 1998) pp.37–49, 226–243.
~~~

4: The Fourth Core Principle
Ecocentric Ethics are Grounded in Awareness of our Place in Nature

Our lives are made difficult and our survival is threatened by the easily detectable truth that most people almost totally lack "a profound appreciation of Earth (which would) prompt ethical behavior toward it." If we seriously seek to behave ethically toward Earth we will have to pursue much more zealously the rightness or wrongness of our actions, the goodness or badness of our decisions, the virtue or vice entailed in our behavior, and the desirability or wisdom of actions, ends, objects, or states of affairs. We must weigh the impact on Earth of all our actions. In spite of the blow to our hitherto unbridled egotism, we are perilously close to multiple catastrophes. These will occur if we stubbornly insist, "Mankind's will must be done." Mark Twain's cryptic comment that "Man is the only animal that blushes, or needs to," should be a recurring subtitle of endless news articles that appear in print to describe the military, economic and personal plights of our times. We are at the point where we must act with more wisdom and restraint than we presently care to employ. Sober reflection on our relationship to Nature, even if never before attempted, might help us to realize that we are as much a part of Nature as any other creature we observe around us.

The world suffers from the sad truth that the majority of our politicians either know very little about the Earth or, in the constant bombardment of human wants, become reconciled to paying attention only to the immediate pressures put upon them. Of course the same is true of voters whose pursuits have long been subject to the goals of an irrational economy. Our survival depends on acting in harmony with natural law. Through education our consciences should be alerted to our relationship with all other living things and with Earth itself.

Industrialists are confused by their erroneous perspective that Nature is the enemy; they need badly to remold themselves to reality. It is an astounding act of arrogance that leaders who think they are on the path of progress should wish to conquer a planet that, in its own magnificence, supports them just as completely as a dog supports the flea on its belly, or a cow supports the warble fly on its back. That "the bottom line is profit" is as serious and eventually lethal a goal, as a tipsy alcoholic's conviction that gin serves as his security blanket. The addiction to profit is a daily reminder that concern for Earth is limited by the symbol of money that is garnered by destroying the health and stability of the ground beneath our feet. Possibly we, as a species, exemplify Euripedes' observation, "Those whom the gods wish to destroy, they first render insane." Unfortunately, as a result of the dominance of politics and industry and their economic philosophy, very little can happen until our leaders awaken from their nightmarish folly.

Control exerted by today's misbegotten power structure even extends to schools, whose curricula are bureaucratically ordained. It is not irrelevant that the *Harvard Law Review* observed that schools are used for indoctrination. Consequently there is no integrated effort to teach students, at any level, that they are Earth products and are genetically connected to other living things. In spite of this arbitrary choice, we and all other living organisms are products of the Earth — born from it and destined to return to it. Neither our religions nor our schools take it upon themselves to protect and cherish the stately grandeur of our planet. If Earth is considered nothing of consequence, and the only important things are a high standard of living in this world and Heaven in the next one, then it is as if we humans were given *carte blanche* to exterminate all other life forms at our convenience and then leave the Earth a garbage dump when we are whisked away to absolute bliss.

It is not difficult to recognize Earth as a creation. Whether Earth happens to have come into being because of a "Big Bang" or a "Let There Be Light" does not really seem to matter. Astrophysicists speak of it in one way, and theologians in another. Personally I find a lot of sensibility in an

Oriental proverb that states, "God will not enquire of thy birth, nor will he ask thee thy creed. Alone He will ask: 'What hast thou done with the land I loaned thee for a season?'"[1]

The Manifesto's authors noted, "Veneration of Earth comes easily with out-of-doors childhood experiences and in adulthood is fostered by living-in-place so that landforms and waterforms, plants and animals, become familiar as neighborly acquaintances." They point out that the rural rather than the urban milieu prepares an individual for comfortable existence in the world of Nature.

Urban Life Brings Estrangement

With roughly half the world's population now living in urban centers, there is increased allegiance to the works of man in preference to the works of Nature. It has become easy for people to assume that the city is the real world and the outdoor places are wildernesses savaged by mosquitoes, black flies and other grim beings, both large and small.

Let's start from where we are. People have become familiar with the term global warming. One of the most reliable forms of evidence is the measurable fact that glaciers are shrinking. Another is the concern that a rise in sea level due to more meltwater entering oceans will cause widespread flooding to many cities and low-lying islands throughout the world.

There is also general knowledge that millions of vehicles exhaust prodigious amounts of carbon dioxide into the air, and many factories do the same. A solution to this problem does lie in the hands of individuals.

On a worldwide basis, production of oil has risen from 68,344,000 barrels per day in the year 2000, to 73,761,000 barrels per day in 2006 (production averaging that daily rate in March 2006.)

According to BP's *Statistical Review of World Energy 2005,* the US "used the rough equivalent of 17.1 billion barrels of oil in energy in 2004, including all conventional sources of energy, from oil to coal to nuclear. That includes about 6.9 billion barrels of oil burned." (*Harper's,* August 2006).

What this adds up to is that we are using incredible amounts of fossil fuel energy with reckless abandon. Much of the energy being used today is mined from algae and other plants that thrived and decayed long ago and were ultimately sequestered in the Earth. Technology has made fossil fuels available for use, but wisdom does inform us that we have acted with great haste in devising trivial uses for this energy.

To be sure, one can see the impact of our economy more clearly if out-of-doors experiences occurred in childhood, and even more so if the individual has remained in a rural atmosphere during adulthood and has witnessed the ravaging of the planet. Unfortunately a life spent largely in an urban setting has served as forcible detachment from the sights and scenes that are part of a person's genetic history. When the city is the norm, one has been effectively estranged from wild places and the more random diversity of rural settings. Nonetheless, some do consider this desirable. The other side of the urban–rural coin is presented in *The Human Zoo* by Dr. Desmond Morris. Morris was Curator of Mammals at the London Zoo in the UK. He draws many parallels between life in zoos and in large cities, the principle difference being that animals in zoos are imprisoned against their will, while hordes of people have chosen urban life or are kept there by the availability of employment.[2]

Biologically, urban conurbations are a form of extreme parasitism on the ecosphere, and they constitute a large footprint (or demand) on the shrinking, supportive, wilder places from which their raw materials are derived. The urban footprint is also enlarged by excessive transportation involved in providing cities with the needed goods created by population density. Additionally, huge, compact populations lack wide, open spaces and thus need expensive centers for recreation and artificial pleasures of all sorts. Our synthetic economic aspirations have caused a dangerous escalation of our numbers and of what has become an unsupportable acceleration of our demands. While we accept the maxim that ignorance of our human law is inexcusable, we have not begun to realize that violation of natural law leads eventually to immutable consequences of an absolute nature.

Our Tribal Origins

Ecosystems (and hence the ecosphere) are known to be quite resilient, but they can also collapse with astounding rapidity. The signs of such collapse are numerous. Millions of individuals have been cajoled and resigned to cheek-by-jowl existence in communities where anonymity is a consequence of excessive population density. Small communities are a reminder that our origin is tribal, and tribes often contained about 250 people. It has been suggested that tribal size was somewhat a measure of the number of people who could know one another quite well. The cancerous growth in importance of business has led to an equally cancerous "bigger is better" mentality. This has eradicated small farms, small communities — and thereby lives — wherein community mores helped young people to grow up amidst familiar faces, landmarks, recreational opportunities, and shared local history. Several generations of family friends and relatives were normally encountered in such places during a single outing or at a community event. Belongingness has been sacrificed to business and the overdone devotion to wealth.

I recall being in Potlatch, Idaho on one occasion and learning that the name was derived from the fact that native tribes held potlatches there. These ceremonies would serve as a model for the extremely constipated greed of our multi-millionaires, billionaires, and others whose focus has been on acquisition. Prior to a native potlatch, new garments and necessities were made for the families in attendance. The potlatch was usually a social event of great significance, for people vied with one another in giving away their most cherished possessions, and who do you suppose gave away their very proudest possessions? As you might expect, chieftains and other leaders, who had the largest numbers of horses, robes, blankets, weapons, and trophies, gave the greatest gifts. To be sure, these potlatches had great social significance and often did mark a special event, such as the birth of a chieftain's son or perhaps the marriage of an important person. Not only were many items given away, but other items were destroyed. The emphasis was on giving away more than one's neighbor. The mass "giving away" and destruction of superfluous goods has sometimes been compared to Thorstein Veblen's *Theory of the Leisure Class.*[3] In this book, Veblen, a Norwegian philosopher who analyzed the "conspicuous consumption" that characterizes modern society, indicated that wealth must be demonstrated and called to the attention of others. Consider the massive effort undergone in preparation for attendance at a modern social, gala event. Is not the preparation for attendance often largely a call for admiration or envy? Does this not bear scrutiny or reflection? It does seem that the Native potlatch demonstrated a finer state of mind through the ceremonious giving away and destruction of hard-won goods. "Getting and giving" is more admirable an ethic than "getting and hoarding."

Do We Substitute Earning for Living?

Excuses will not suffice. It is very likely true that being brought up amid the sights and sounds of Nature constituted a more natural experience than maturing in the center of a city. But there are other ways that one can learn vicariously. Quite possibly much of today's stress results from the feeling that we are leading meaningless lives. Nothing stops us from finding a quieter place where one can reflect and view some of the awesome wonders of Nature. Many of us need a change of pace, a reprieve from the endless "chase for nothing" that is proliferated by the mass media. The fewer things we need, the richer we are. We have lost sight of the truth that our legs were made for walking. An interesting example of this appeared in Thoreau's *Walden:*

> One says to me, 'I wonder that you do not lay up money; you love to travel, you might take the cars and go to Fitchburg today and see the country.' But I am wiser than that; I have learned that the swiftest traveler is he that goes afoot. I say to my friend, suppose we try who will get there first. The distance is thirty miles; the fare ninety cents. That is almost a day's wages. I remember when wages were sixty cents a day for laborers on this very road. Well, I start now on foot and get there before night; I have traveled at that rate by the week together. You will in the meantime have earned your fare, and arrive there tomorrow, or possibly this evening, if you are lucky enough to get a job in season. Instead of going to Fitchburg, you will be working here the greater part of the day. And so, if the railroad reached round the world, I think that I should keep ahead of you; and as for seeing the country and getting experience of that kind, I should have to cut your acquaintance altogether.[4]

And later, in reflection on life, Thoreau wrote: "This spending of the best part of one's life earning money in order to enjoy a questionable liberty during the least valuable part of it reminds me of the Englishman who went to India to make a fortune first, in order that he might return to England and live the life of a poet. He should have gone up garret at once. 'What!' exclaim a million Irishmen starting up from all the shanties in the land, 'is not this railroad which we have built a good thing?' Yes, I answer, comparatively good, that is, you might have done worse; but I wish, as you are brothers of mine, that you could have spent your life better than digging in the dirt."[5]

Bringing this more up to date, Grant MacEwan wrote, in an essay entitled "Loaned For A Season," "To enjoy the full benefit of an exalted position, he (man) plundered soil, grass, trees, minerals, wildlife, and other natural gifts, forgetting that they were his 'for a season' only. Knowing his life span was short, haste seemed imperative. That succeeding generations with expanding populations would have more rather than less need, conveniently escaped his notice."[6]

I remember when I was teaching, talking to some students one day when having lunch. One of them was expressing his admiration for a Cadillac he had seen in a showroom window and stated his intent to own one some day. Several others disagreed and thought it would take too long to earn $30,000, the price at that time. I was unexpectedly asked my opinion. I owned a bobtail jeep at the time, because I liked to go up back roads to roam in the woods on weekends. I gave as honest an opinion as I could and reminded the students that the real price of an item was the time you had to spend working to pay for it. I expressed my opinion that "life time" is a limited good, and the more things you had to buy, the less time you would have to do the things you wanted to do. Quite noticeably we continuously trade off "life time" for possessions we feel we must have, and buy one wished-for item after another, spending our years in that manner.

The Consumer Trap

We are reluctant to face the need for change, although many people are aware that we are enmeshed in a tightening noose of our own making. We have been trained to be consumers and are titillated by new gadgetry that appears on the market with impressive regularity. Yet, we occasionally do realize that we are overburdened with possessions, and with increasingly large debts.

"A Manifesto for Earth" recognizes that attachment to Planet Earth is more easily felt if we grow up in natural surroundings and live in small communities. Whether or not we have attachment from familiarity with Nature does not alter the fact that an urgent change of behavior is needed now. We must respond to world problems with conservative behavior that presently seems virtually unthinkable. We pass the buck and wait for politicians to solve our problems for us. They, themselves, founder in unwillingness to face the facts, with the result that their proclamations are vague, insincere, and often circumlocutory. They are unfortunately shackled to the reigning politico-industrial colossus, which is horribly blinded by its own addiction to greed, wealth, and reductionism. What a sad situation that an entire society is so habituated to its material cravings and routine that it hasn't the courage or insight to clean up its act!

Let's take issue with some of the unthinkable problems. One of them, certainly, is our addiction to those four-wheeled, automated, multi-passenger wheelchairs that we call automobiles. As well as millions of pages of advertisements used to promote their sale (paper-wasters), automobiles are a hideous waste of energy and a vastly successful money-trap for producers and for the petrochemical industry.

Looking back some years, there was better mass transportation available than now exists. There were passenger trains that arrived and departed with stunning punctuality. There were buses and trolleys in many locations. Unfortunately there were also corporations that could foresee and arrange the slavish attachment of people to machines. So they bought up and rendered unusable these means of mass transportation. The availability of gasoline and of Henry Ford's mass-production led to the popularity of automobiles. Ford then sold automobiles in any color you wanted, "as long as it is black." Though farmers and many townsmen labeled cars "stinkbuggies," the market possibilities were so promising that cars first became fads and later necessities. As the regression we call progress continued, corner grocery stores also disappeared and large central markets, which necessitated use of autos, became popular until they were succeeded by shopping malls. "Keep them on the move. Keep them buying gasoline." The result of a few generations of servitude to automobiles is that we are now incorrigibly addicted to them and willingly suffer enormous costs and an increasing drain on our personal resources because of them. Strangely enough, we are not allowed to deduct their cost on our income taxes. The easiest thing to do is to chortle, "You can't go back; we are living in the best of all possible worlds." Addicts to alcohol, psychedelic drugs, incessant travel, speed for the sake of speed, and travel for the sake of travel, are too thoroughly on the hook to wriggle off it.

The Incredible Appetite of Aircraft

A vast pollution problem has also resulted from intensive use of aircraft. The observation was made on a radio broadcast that at any given time there are more than 30,000 aircraft in flight. Certainly they facilitate travel. Unfortunately they are superlative pollution machines. An article in the magazine *ecologist* (June 2005) quoted (from the *Guardian*) an article written by George Monbiot on the subject of a proposed wind farm at Winash in Cambria. "The Winash project, by replacing energy generation from power stations burning fossil fuel, will reduce carbon dioxide emission by 178,000 tonnes a year. This is impressive, until you discover that a single jumbo jet flying from London to Miami and back every day releases the climate-change equivalent of 520,000 tonnes of carbon dioxide a year. One daily connection between Britain and Florida costs three giant wind farms."[7]

The foregoing speaks for itself — or at least should. The Manifesto states, "An awareness of self as an ecological being, fed by water and other organisms, and as a deep-air animal living at the productive, sun-warmed interface where atmosphere meets land, brings a sense of connectedness and reverence for the abundance and vitality of sustaining Nature." Today's problems center on the fact that we do not actively realize our connection to Earth, but instead revel in the technological monstrosities that have led us to the brink of self-destruction. We are so enamored with machines and power that we refuse to evaluate their impact on fragile and intricate living systems.

It is of course true that if you have the awareness of which the Manifesto speaks, it does bring about a deep-seated realization of our relationship to our planet, and an abiding loyalty toward it. This probably accounts for some resistance toward endless servitude to machines. Unfortunately, inadequate education at home and at school, plus continuous trivial and sensational, but superficial, information from media steadily feeds the public on business blarney designed to keep them in the role of obedient consumers. There is yet much truth in an observation more than a century old: "Blessed be the man who never reads a newspaper, for he may see Nature and through her the Divine."[8]

Regaining Individuality

So what can we do? Or even, what must we do? There are two things that are essential: We must regain our individuality and we must reduce our alienation from Nature.

In regard to regaining individuality, I can provide a mild example that I found interesting when it occurred. A visitor one day, after more casual conversation, made the comment, "It doesn't look as though you have a television."

I replied, somewhat neutrally; "We both enjoy reading and prefer doing that to watching TV; just a matter of choice I guess."

He then ventured, "I don't have TV either, and I will tell you an interesting experience I had. I was in a furniture store with my wife when she was shopping. There was a sale on televisions, and a salesman collared me when I glanced at a TV while I was waiting. The salesman, of course, wanted to sell me a set and asked me how old my television was. When I told him I didn't have one, he really got his sales pitch into high gear. After I listened to him for a bit I took a different tack and asked him how much he would pay me to take one of them.

"I caught him off guard and he just stared at me while I told him that people should be paid to take one home. I explained to him that I would have to provide it with space that could be used for other things, that I would expect to be paid to watch the advertising promotions, and that the seller should cover my electricity bill. I also told him I found TV offensive. I wasn't interested in two minutes of any program followed by endless advertising, and so on. I think that maybe he figured I was a little around the bend, but I also suspect that I gave him a new perspective. Anyway he remembered me, and when I was in the store a year or so later, he asked me with a grin if I had bought a TV yet, and I told him that I had not."

One need not go the length my visitor took, but we should realize that we have become captive audiences for all sorts of advertising; the promoters of products have carefully studied the effect of repeated advertising of brand names and propaganda on the subconscious mind of purchasers. There are far more valuable ways in which time can be spent than by merely being an at-home spectator.

I will admit that the foregoing bit about television may seem rather trivial, but there do seem to be many people whose evenings have fled away from them because it is easier to sit back and be amused by a machine than to undertake reading a book, studying something of interest, or even devoting oneself to reflection about the life one is living.

Plato's Cave

Plato presented the classic example of the illusion under which humanity toils in the seventh book of *The Republic.* Plato's fame has earned numerous approbations, such as British philosopher Alfred North Whitehead's comment that, "All subsequent philosophy is but footnotes to Plato." Whitehead of course could not foresee the emergence of an Earth-focused, "ecocentric" worldview. In *The Republic,* Plato speaking through the voice of Socrates, depicts the human situation as one lived within a cave, where the inhabitants are focusing upon an unreal shadow world that guides the senses to a false interpretation of life. In the words of Socrates:

> Behold! Human beings living in an underground cave, which has a mouth open towards the light and stretching all along the den; here they have been from their childhood, and have their legs and necks chained so that they cannot move, and can only see before them, being prevented by the chains from turning round their heads. Above and behind them a fire is blazing at a distance, and between the fire and the prisoners there is a raised way, and you will see if you look, a low wall built along the way, like the screen which marionette players have in front of them, over which they show their puppets.[9]

Socrates goes on to speak of the people passing behind the low wall, carrying all sorts of things such as vessels, statues, animal figures, and various other items that project above the wall. The fettered people in the cave see only their own shadows, and the shadows of the things being carried by the people behind the low wall. They also hear, echoing from the wall in front of them, those things that are said by the passers-by.

To these people, Socrates said, "The truth would be nothing but the shadow of the images."

Socrates goes on to say that if a prisoner were suddenly released and compelled to leave the cave and go out into the light, he would first be blinded by the glare and would be unable to realize that what he was seeing was reality rather than illusion. He would be perplexed and would prefer, after only a cursory look at things in the light, to feel that reality was the shadow world of the cave. As Socrates points out, it would take time for an individual to become accustomed to the sight of the upper world, or to be able to emerge into the light of the sun and to withstand its brightness.

Eventually though, the freed individual would see things clearly in sunlight and would think with pity of those still in the cave. He might return, especially if asked, to try to help those in the cave to understand their true situation.

Socrates explains what the attitude of the people in the cave would be upon the return of their would-be-guide.

> Now if he should be required to contend with these perpetual prisoners in 'evaluating' these shadows while his vision was still dim and before his eyes were accustomed to the dark — and the time required for habituation would not be very short — would he not provoke laughter, and would it not be said of him that he had returned from his journey aloft with his eyes ruined and that it was not worth while even to attempt the ascent? And if it were possible to lay hands on and to kill the man who tried to release them and lead them up, would they not kill him?[10]

Obviously Plato saw humans as individuals entrapped by their passions and desires. Plato's conclusion was that humanity lives backward and does not pay attention to the real events taking place and the effect they have. It is not surprising that philosophers and theologians of our own time have expressed great concern about the dominance of "technical knowledge" over our lack of spiritual concern, and of the shallow base we have therefore created for moral and ethical thought. Philosopher Paul Tillich stated, "the flowering of technological civilization has taken place at the price of loss of depth in humanity."[11]

Wisdom and Technology

Reflect, too, that Plato, 2300 years ago, realized and emphasized the superiority of *philosophia* (the love of wisdom) over *techne,* which referred to technological or manipulative knowledge. In our own time there has been massive subordination of wisdom — not just of wisdom, but also of the reflective, slower pace of life that might enable people to have time to ponder just where all this mayhem we call "progress" is really taking us. Additionally, a changed pace of life would enable people to be much more aware of the negative, and seriously dangerous consequences of unbridled growth. Lacking moral restraint, science has made possible not just useful things, but other things very destructive to all life on Earth. Distorting the atmosphere with insane amounts of carbon dioxide, arranging a "global economy" that leads to food traveling thousands of miles before it appears on tables, building nuclear and biological weapons not just to prevent, but more to guarantee worldwide Armageddon, manufacturing carcinogenic biocides and other life-threatening industrial products; these are not acts of responsible leaders.

Just what is the root of the problem? Quite possibly the answer was given 600 years before the time of Christ. More specifically, it was in the time of Confucius. Known as China's most prominent teacher, with more than 3,000 pupils, Confucius was also famed for his administrative skill, as well as for his refusal to compromise his integrity. He is well remembered as a visionary idealist. Although revered, political leaders feared him. Once when he was considered for the leadership of Lu (his native state), he was asked what would be the first thing he would do. He replied that he would call for a "rectification of names." He explained that if a man was called a prince, he should actually be a prince in every sense of the word, a man whose life and habits, whose moral stature and trustworthiness would meet in every way what would be expected of a prince who was truly noble to the core.[12] There is no question that our world is awash in people given a dignified title of some sort. We do not suffer for lack of His or Her Worships, for the Honorable this and that, for military officers who nearly choke on their gold braid. However many titles have been granted, we are still wrecking the planet. We apparently have totally overlooked an essential consideration, namely that we are often rewarding behavior that helps erode, rather than stabilize, our relationship with the Earth itself. This possibly rests on narrow homocentricity that ignores the most important relationship of all, which, in the long run, is our true relation to the ground beneath our feet. The ridiculousness of the human search for power and glory is a single example of alienation from reality, an anomie that would be dispelled if we strive to become dutiful and diligent citizens of Planet Earth.

It is somehow ironic that our modern "elite," if that name suffered rectification, might be described as perverted by greed and ambition, and also as obsessed with the belief that endless wealth and power are the goal of all life, which entitles them to be veritable gods on our abused planet. It is paradoxical that the things for which people expect renown are often the very things that merit disgrace. Plato's contention that we live backward deserves reflection.

A rectification of names would be a good idea, and might be a mark of maturation, sorely needed by our species.

Historian Arnold Toynbee deduced that the two leading pursuits of governments are making war, and controlling the public. War is actually a breach of the responsibility that politicians have toward their electorate, and toward the Earth. Continuous explosions, craters from bombing, extensive and expensive damage to cities — now that the military has prevailed on political powers to allow the bombing of civilians — all of these effects and many others, indicate that the Earth itself can no longer afford wars and that our leaders are lacking the wisdom needed to keep the planet healthy enough to sustain life. It is far too easy for a man in office to order others to lose their lives because of the inability of governments to govern in such a manner as to make peace on Earth become a reality.

Reality Beckons!

When I have a few minutes of leisure I frequently pick up my copy of *Meditations* by Marcus Aurelius. These are the thoughts of a Roman emperor and general who camped along the Rhine to discourage invasion by Goths. From what I have read, he preferred the company of his troops to ceremonious life in Rome. I came across this bit recently: "Soon Earth will cover us all. Then in time Earth too, will change. What later issues from this change will itself in turn incessantly change, and so again will all that then takes its place, even unto the world's end. To let the mind dwell on these swiftly rolling billows of change and transformation is to know a contempt for all things mortal. How ignoble are the little men who play at politics and persuade themselves that they are acting in the true spirit of philosophy. Babes, incapable of wiping their noses! What then, you who are a man? Why, do what nature is asking of you at this moment. Set about it as the opportunity offers, and no glancing around to see if you are observed."[13]

So, is there something that Nature is "asking of you at this moment?" I think there is. Nature, I would say, is asking us to give up our alienation from Earth, and to reject the economic seduction in which we are allowing ourselves to be drowned.

We all do stand at some sort of inevitable turning point with, as I see it, no alternative other than working diligently, even denying ourselves our delight in personal mobility for the sake of peace on Earth and good will toward all living things. If we continue to worship wealth and power, we will disappear with a bang, a whimper, or some other form of eradication. If the planet can recover from the increasing temperatures we have guaranteed for some years, our disappearance would perhaps be a blessing. That's sad, because we should not have given up the future of our species through choosing our Faustian search for unlimited energy and for the gadgetry to which we pay daily obeisance.

We should no longer kid ourselves. Progress does not consist of running behind the machine. Said machine has been out of control for years. True progress will consist of shutting down the machine, of expressing loyalty to the planet and of reshaping our activities and expectations in such manner. Insofar as possible, the bludgeoning and devastation we have inflicted on the ecosphere must be remedied. This will require intelligence and sacrifice. Although we will have difficulty addressing many of the problems we have created, we have truly run out of other options. The society that was formed years ago, and called itself Earth First, identified changes we must make and problems that can only be solved if our loyalty to the planet is unswerving, dedicated, and scrupulously intelligent. We must put Earth first.

[1] *The Best of Grant MacEwan,* p.58.
[2] Desmond Morris, *The Human Zoo* (Toronto: Bantam Books of Canada Ltd., 1971) p.11.
[3] Thorstein Veblen, *Theory of the Leisure Class* (New York: The Viking Press, 1931) p.75.
[4] Henry David Thoreau, *Walden* (Toronto: Random House, Inc., 1950) p.47.
[5] *Walden,* p.48.
[6] *The Best of Grant MacEwan,* p.58.
[7] Jeremy Smith, "Green Electricty…Are You Being Conned?" *ecologist*, June 2005, p.58.
[8] Walden
[9] *Great Books of the Western World,* Vol. VII, "The Republic," p.388.
[10] Ibid.
[11] Robert Earl Cushman, *Therapeia* (North Carolina: Univ. of North Carolina Press, 1958) Ch. 13–16, p.xvi.
[12] Will Durant, *Our Oriental Heritage* (New York: Simon and Schuster, 1954) p.666.
[13] Marcus Aurelius, *Meditations* (Great Britain: Penguin Books, Ltd., 1975) p.78.

5: The Fifth Core Principle
An Ecocentric Worldview Values Diversity of Ecosystems and Cultures

Geology suggests that developments in ecodiversity and biodiversity were strongly affected some 200 million years ago by the breaking apart of an original supercontinent called Pangaea, which began to split into the continents that we know today. The theory of continental drift was proposed by the German meteorologist Alfred Wegener who, in 1915, discovered that plants grown in Greenland still showed similarities to ones grown in tropical areas, and that strong evidence of glaciation can still be

noted in tropical countries such as Brazil and Africa. Gradually, Earth scientists found other evidence of such major changes during geologic history. Studies of ancient mountains revealed that the Appalachian Mountains, which extend from the eastern United States into Newfoundland, were likely once connected to the Caledonian mountain system located in Scotland, Northern Ireland, and Scandinavia.

Paleontologists have discovered fossils of similar land mammals in rocks a hundred million years old in Asia, Europe, and North America. Later use of radioactive dating of rocks identified fossils similar in age and type in Africa and South America. Studies that formulated the concept we know now as plate theory had indicated by the 1960s that as many as sixteen crustal plates exist beneath the Earth and movement of these plates brings about significant geological change through the pressure they exert upon one another.

An interesting influence this has had on biodiversity is that while different grazing animals exist in different continents, and are preyed upon by different predators in each location, the same ecological niches are filled on each of the continents. Thus, big catlike animals such as lions in Africa, cougars and jaguars in North America, and Siberian tigers in Russia are major predators that very likely developed adaptations to fit the environmental conditions in which they lived.

It has also been noted that both a marsupial wolf and a marsupial tiger once lived in Australia. In fact they lived there until not too many years ago, and their extinction was accelerated by the introduction of predators from other continents. Why would such introduced predators have affected the Australian marsupial so severely? Part of the answer is that the young of marsupials spend a considerable time being carried around in the mother's marsupial pouch (marsupium). As the mother marsupial tiger or wolf was impeded by carrying her young around in her pouch, she was no match for the more active predators that reproduced by giving live birth to young that could range with the mother or be left in a safe place.

Australia still has two kinds of mammals (Monotremes) that are hatched from eggs. These are the platypus and the echidna. Both animals serve as significant reminders that as mammals evolved they moved from egg-laying in their more rudimentary state through a marsupial state in which, though live young were produced, they required further development and nutrition in the marsupium before being able to function independently. Unlike placental mammals, they brought restricted movement upon their mothers, and the young remained totally dependent for a longer while after birth.

The survival of the echidna and the platypus is probably accounted for because the ecological niches occupied by each of them was quite specific and did not cause a serious conflict with other species that occupied the same niches.

The echidna is also known as the spiny anteater. Its strong spines cloak the hairy body. Its mouth develops into a horny beak and it eats ants, which are captured easily by a sticky tongue and brought into the echidna's toothless mouth. The mother echidna lays a single egg, and when the egg hatches the young mammal is nourished from the mammary gland within the mother's pouch and remains there until its growing spines signal the mother that it is time for the young one to go off on its own.

The platypus, also known as the duckbill, is a small and unusual mammal. It inhabits cold streams and tropical swamps of eastern Australia and Tasmania. The duckbill has a flat beaver-like tail and webbed feet that it uses for swimming and digging burrows. Its muzzle is bill-like and sifts insect larvae, small crustaceans, and mollusks from the water. The food taken in is stored in cheek pouches until swallowed. The young are carried in the marsupium and feed by lapping up milk secreted from the mammary glands. Duckbills are covered by fur and are recognizable as mammals. Both the echidna and platypus are unique examples of the specializations that may result from biodiversity.

Life Forms Differ In Needs

The continental drift theory further substantiates the likelihood that varying life forms have been affected in their development by great differences in soils, waters, and by local climates. Plants find suitable habitats in soils of varying acidity, and particular plants often have specific mineral needs that must be met if they are to grow and thrive. Animals also have unique needs, but their mobility enables them to travel lengthy distances to salt or mineral licks which they may visit at intervals. Study of species has determined the maximum heat or cold that a certain species can tolerate, and species vary between those that function best in narrow ranges of temperature variation and others that can adapt to wide temperature ranges. Most people are aware that some animals, such as varying hares, change the color of their pelage seasonally, becoming white in winter as a natural winter camouflage and turning to a tan or brown color in summer. Weasels also display similar seasonal shifts in color.

Another adaptation to seasonal changes is known as altitudinal migration. Species such as mule deer will graze in mountain meadows in summertime and as fall chill brings snow to high elevations they will migrate downhill to areas with less snow. Other species, such as grizzlies, range in high country (about 8,000 feet in southern British Columbia) where their favorite prey, hoary marmots, are to be found. They often hibernate in the mountains, turning in early for winter. When they awaken in spring they usually move downhill to elevations where succulent vegetation is available earlier. Other high country animals that remain active all year are more often seen in winter at lower elevations. These may be predators. For example, cougars follow deer downhill, and weasel family members such as marten, fisher, and wolverine are more common in winter in lowlands. Note that not only is there diversity of species, but also diversity of behavior that results from adaptations to seasonal change and the migration of food species.

Biodiversity also has to do with intelligence in animals, something easily observed by country residents and quite often denied by persons more remote from opportunities to observe animal behavior. In *Intelligence in Nature,* Jerome Narby commented on an interesting example of storage and recovery of seeds by Clark's nutcrackers, an alpine bird related to crows and ravens. Quoting a study by Kamil and Balda in 1985, the author wrote, "Nutcrackers expend substantial amounts of time and energy during the late summer and fall harvesting seeds from pine cones, transporting them up to 22 km., and then burying the pine seeds in thousands of discrete caches."[1] These seeds, which may number well over 20,000 in a single season are recovered in winter and spring, and permit nutcrackers to winter and breed early in alpine habitats where food is not otherwise abundant. Field observations have discovered that nutcrackers have high spatial intelligence and are capable of recovering seeds cached months before. This does constitute a remarkable adaptation to stringent climatic conditions.

Extinctions Caused by Humans

Regarding biodiversity, it has been noted that human settlement in such places as New Zealand, Madagascar, and Australia, led to rapid destruction and eventual extinction of mega fauna. Among other species destroyed in Madagascar were a number of large and flightless elephant birds, and seven genera of lemurs, which were the primates most closely related to apes and humans. Edward O. Wilson, a Harvard university scientist commented in *The Diversity of Life* that the species rendered extinct included a species of lemur that ran on all fours, and another as big as a gorilla, that climbed trees. He also mentioned "an aardvark, a pygmy hippopotamus, and two highland tortoises" as other species that were destroyed.[2]

Wilson also refers to the extinction of thirteen species of the large flightless birds called moas that were almost totally wiped out by Maoris before 1800. These people arrived in New Zealand about 1000 AD. Australia, where humans arrived about 30,000 years ago, lost many species of large animals, including marsupial lions, a species of kangaroos about eight feet (2.5 meters) tall, and other species including animals that resembled rhinoceroses, tapirs, and ground sloths.[3]

Listing some of the animals rendered extinct directly by humans or through human effect on habitats provides an example of why we should avoid further destruction of remaining Earth fauna. There are people who may choose to be quite chauvinistic, and regard human destruction of species as within our rights. This disregard for other animals overlooks the roles of animals such as coyotes, weasels, hawks, and owls that are important predators on mice and voles, and if unchecked, can rapidly achieve plague populations.

Is it not a paradox that our disdain for other forms of life has long been an infamous dereliction of the deepest meanings of the words "humane" or "humanity"?

Analysis has been attempted of the mass of extinctions that took place at the end of the last ice age some 11,000 years ago. The book *Quaternary Extinctions,* edited by Paul S. Martin and Richard G. Klein, would be of interest to individuals who wish to peruse this topic extensively.

Considering the impact that our species has had on animal populations in many regions newly discovered by expeditions or in areas newly populated by people, an important characteristic of humans is revealed.

This is a tendency to immediately exploit whatever is discovered. To early European settlers the forests of North America seemed boundless, the amount of fish in streams prodigious, the quantity of bison inexhaustible, the number of passenger pigeons so limitless that hundreds of thousands were shot when roosting in order to feed pigs driven in among the corpses. The flaw in human nature is immediately obvious and is epitomized by forest industry managers who choose to ignore the major importance of forests in climate stabilization and other ecological benefits. One of these managers, with an extremely homocentric attitude regarding trees in his area, stated, "It's ours, it's out there, and we want it all. Now."[4] Modern developers, who have convinced themselves that turning pristine acres into revenue sources of any kind is a mark of progress, frequently exhibit the same attitude. Note carefully that such a narrow definition of progress results in de-development of Nature's millions of years of balanced, vibrant, and dynamic stability. In its place such individuals happily substitute whatever monocultural monstrosity they elect, whether it is a subdivision, a wheat field, or a parking lot fancied by their project. Strangely enough, when man destroys the work of man, be it machine, house, artwork, or statue, he is called a vandal; but when he destroys a moose for its antlers, a bear for its hide, or a fish to be mounted to display his prowess as a trophy, he is called a sportsman; and if he destroys natural landscape for an oil well, a subdivision, or a casino, he is called a developer.

What is the problem? It is that short term values, as short as those of any other predator, dominate humankind. As a species, immediate satisfaction negates all other drives. The goal is food on the table today and wealth in the pocketbook without delay. Now that freezers are available, many of us know of individuals who have two or three freezers full of food, who still go a-moose hunting, or go fishing even for spawning fish, when the only threat to their food hoard is a power failure. It is quite obvious that a conservation ethic is largely non-existent, and many are belligerent that anyone should dare to propose such an ethic.

Various biotic factors, such as the availability of needed foods and threat from predators, also affect the survival of a species. This year, hot, dry weather has enabled a bumper crop of ruffed grouse to survive. In the fall, hunters will easily secure grouse for the table. Also, Swainson's hawks, which prey on grouse in early and mid winter, will have good hunting. On our winter snowshoe walks we frequently come upon rabbit or grouse tracks that end abruptly. The evidence of their fate is presented by wing marks in the snow that explain the reason for disappearance of the tracks.

Subtle physical factors come into play. Slope faces (aspect) may determine the extent of incoming heat, wind patterns, and the rate of evaporation; all affect the suitability of the environment for particular species. And, for some species, there must be particular microclimates. In hot, dry country, slopes that face the sun may be mainly or completely covered with grass, while cooler, north-facing slopes with milder temperatures and reduced evaporation may be covered with timber.

Cultural Diversity

It is natural that the effects of physical ecodiversity and accompanying biodiversity also affected the cultural diversity of people spread across the globe. Sober reflection will make us realize that the cause of diversity among people has been spawned from the possibilities offered by ecosystems in which they live. Habits, beliefs, thoughts, traditions, and language stem from the biodiversity of the landscape, from the climate in which an individual is born, from the unique conditions of a particular habitat, and ultimately from the expectations and taboos of the group in which one becomes a member.

For example, if an individual's place of birth is along a coastline or on an island, and if products of the sea are the most available food choice, then one responds by developing a culture dependent on the ecosystem benefits most prevalent. For indigenous Haida in the Queen Charlotte Islands of British Columbia for example, clothing and longhouses were largely made from cedar trees. The bark was fashioned into clothing, while the logs were crafted into sturdy dwellings and canoes. Cedar bark became useful for making lines for fishing and cords for nets. Suitable rocks would be pierced so that a fish line could be passed through them, thus providing a sinker to carry the line down for bottom feeding fish. The sinkers could be further developed as works of art. The development of totem poles from cedar could also have resulted from the workability of the wood. Also, with abundant food available from the sea, craftsmen had the time to develop artistic skills. Totem poles are thus a hallmark of the expressiveness of individuals. They were fostered by thought and experience enhanced by superstitions or religious impulses arising from tribal cosmology. One can also understand the development of warlike attitudes arising from species territorialism, from competition for food, and from the predatory instinct further stimulated by opportunity, or from the quest for loot, or slaves.

It is not difficult to envision various types of culture arising in prairie, semi-desert, uplands, arctic, or other ecological settings. Cultures have arisen from the availability of a food species such as bison. This migratory species also compelled migration of other species, including humans who preyed on them. Abundant populations of pronghorn antelopes, prairie chickens, and migratory birds further enriched this particular food source. For some tribal groups, horses became part of the culture. Note too, that migratory habits among people were fashioned by not only shifting food supply, but also by the winter need for shelter from fierce prairie winds and drifting snow.

One may also note the cultural diversity fostered by domestication of animals, by development of trade leading to mule trains or camel caravans in some locations. Also significant in human history is the more than 10,000-year-old pursuit of whales for food, for oil, baleen, and even for the use of whalebones to support shelters. These uses ultimately led to the pursuit of whales from pole to pole, and to near extinction. We are hardly aware any longer that whale oil was widely used for street lighting before petroleum was discovered and utilized.

Consider also what the cultural impact has been from the Chinese development of firecrackers, the later development of dynamite for activities such as mining and road building, and even the origin of the Nobel Prize that indirectly grew from the discovery of explosives. Realize, too, that the further work on energy and matter by Einstein led to the nuclear bomb as a next step, and later to nuclear energy plants that provide electrical energy. There are also the later nuclear events at Chernobyl and Three Mile Island to remind us that the loss of control over our energy choices can be disastrous. And of course there is the still continuing effect of expansion of the firecracker story. The nuclear-war rattling of George Bush and Company is now being exacerbated by an unknown number of other nations edgily girding their loins for an unwanted (?) "super war" that could arise from the belligerence of vain politicians whose incompetence is exceeded only by their rhetorical clamor.

For all we know, microorganisms may be called upon once more to develop a multicellular being that has enough compassion to modify its intellectual instability. Perhaps this task has been theirs more than once.

We stand in immense need, for our survival depends on development of a world of peace, compassion and intelligent attempts to live in harmony with Nature. While we are a long way from these goals, we must begin with a great respect for the entire wholeness of which we form such a potentially great or ominous part.

Cultural diversity arises, at least in part, from early behavior learned as a result of need. While members of tribes in early years no doubt suffered or died at times from the need to utilize risky food sources, the culture of the tribe eventually included many safe, alternative food choices. Indeed their food choices were probably safer than those offered to supermarket buyers of today who must contend with toxic additives in the form of pesticides, heavy metals, dangerous preservatives, and unknown genetically modified foods that may have had their origin in amateur chemistry sets once offered as toys.

Inventions for Invention's Sake

Protective measures, such as the use of fire, the retreat to caves, or the development of weapons including stones, clubs, bows and arrows, made people more capable of defense or of hunting other creatures as food. Today's residues of these necessities have unfortunately triggered unbridled technological innovations that have moved far away from protective or life-sustaining measures. Quite blindly we have launched too many inventions simply because we can propagandize them into items that have high marketability. This indicates our harum-scarum nature, our addiction to anything that will produce wealth, and our severe lack of sufficient wisdom to control technology. When Bill McKibben, in *The End of Nature,* lamented that we may not be much smarter than beavers, he was, in effect, reminding us of our relationship to other organisms.[5] All living things, including the tiniest bacteria, are made up of proteins encoded in DNA and RNA molecules. The relationship between species includes astounding similarities as well as unique differences. We choose to focus on the differences and ignore the things we have in common with species we dismiss as inferior to us. It is worth reflecting upon the fact that 99 percent of the genes of mice are also present in people. The general greedy pursuit of material goods suggests that humans also share a lot of pack rat behavior, exemplified by our ritually frenzied, acquisitive habits.

Culture did not stop at the simplest needs, of course. People learned to use horses, thereby increasing their speed of travel and relieving themselves from carrying large burdens. Eventually they learned to substitute horsepower under automobile hoods for the more eco-friendly horse and buggy form of travel used by our ancestors.

Tools have been a substantial part of human culture, and have been vastly improved over time. Logs that could be rolled suggested wheels. Inclined planes could be used to raise heavy weights with less effort. Digging tools moved from hands and nails provided by nature, to sticks and stone tools for digging and plowing. In due course steam engines began our move toward the powerful technologies we have today.

Culture of course has spread through many areas of experience. It has to do with knowledge, with law, art, morals, dress, religions, birth, death, and marriage ceremonies, with traditional dances of supplication to the gods and to the elements, with clothing styles that range from sarongs in tropical climates to fur parkas in the arctic. Many useful projects for students could enhance their understanding of cultural diversity. Study of various animals used as beasts of burden, of dwelling styles, and of traditions could be identified with different cultures. History of religious cultures could be compared. Bowing, shaking hands, embracing, kneeling, removing head covers, and many other cultural behaviors exist in the world and are interesting material for study. For instance, Westerners are taught that an honest person will "look you right in the eye," while some Asian cultures consider such behavior be to be rude.

There is a significant poem to be found in *Farmers of Forty Centuries* by Dr. F.H. King, which strikingly illustrates cultural diversity, and would provide spirited discussions material in the classroom:

> Two little maids I've heard of, each with a pretty taste,
> Who had two little rooms to fix and not an hour to waste.
> Eight thousand miles apart they lived, yet on the selfsame day
>
> The one in Nikko's narrow streets, the other on Broadway,
> They started out, each happy maid her heart's desire to find,
> And her own dear room to furnish just according to her mind.
>
> When Alice went a-shopping, she bought a bed of brass,
> A bureau and some chairs and things and such a lovely glass

To reflect her little figure – with two candle brackets near –
And a little dressing table that she said was simply dear!
A book shelf low to hold her books, a little china rack.
And then of course, a bureau set and lots of bric-a-brac;
A dainty little escritoire, with fixings all her own
And just for her convenience, too, a little telephone.
Some oriental rugs she got, and curtains of madras,
And then a couch, a lovely one, with cushions soft to crush,
And forty pillows, more or less, of linen, silk and plush;
Of all the ornaments besides I couldn't tell the half,
But wherever there was nothing else, she stuck a photograph.
And then, when all was finished, she sighed a little sigh,
And looked about with just a shade of sadness in her eye:
'For it needs a statuette or so – a fern – a silver stork –
Oh, something, just to fill it up!' said Alice of New York.

When little Oumi of Japan went shopping, pitapat,
She bought a fan of paper and a little sleeping mat;
She set beside the window a lily in a vase,
And looked about with more than doubt upon her pretty face:
'For really – don't you think so? – with the lily and the fan
It's a little overcrowded!' said Oumi of Japan
— Margaret Johnson in *St. Nicholas Magazine*[6]

Cultural Behavior in Other Animals

Are other animals excluded from cultural behavior? Not so! Elephants are known to break off tree branches, hold them in their trunks and whisk them around to keep off flies. In the Asian tsunami on December 26, 2004, an interesting tale was told of an elephant that was responsible for saving the people of a village. The village in question had a pet elephant, which was used for hauling logs, and was very well behaved with children. Shortly before the tsunami struck, the elephant acted in an alarming manner. It swept several of the village children up in its trunk and headed for the hills. Alarmed, the villagers followed the elephant trying to make it stop and release the children. Later the villagers realized that the elephant was able to foresee the coming wave, and those people who followed it to high ground were saved while their village was demolished by the enormous wave!

Chimpanzees catch termites by peeling a twig and sticking the wet, sappy end into a termite's nest. When the twig is pulled out, termites are stuck to it — an innovative way of catching food! Woodpecker finches likewise use a toothpick-like twig to spear food morsels. Honeyguides, birds that live in Africa, live mainly on beeswax. They cannot get at the honey themselves, so when they discover a beehive they call excitedly and thereby attract mammals that eat honey. These animals break into the hive and eat the honey. The honeyguide afterward feeds on the beeswax and on the remaining bee larvae. No one is certain how this behavioral means of getting food began. Nature of course has many secrets.

In his interesting book *Animal Days,* Desmond Morris describes an experiment in which he introduced art material to a chimpanzee named Congo. One day he gave a pencil to Congo and rested the pencil point on a piece of cardboard. The chimp moved his arm a bit and noticed the mark left by the pencil. Congo stopped and stared, moved the pencil some more and made a series of lines. Congo began to draw and exhibited a great interest in this pastime although he never made a representational drawing as of a face, another animal or a specific object. Many individuals would be interested in reading about the events of Congo's art career as written in *Animal Days.* Over a period of two years, the artistic chimpanzee made 384 drawings, and later paintings. The astounding part of the story is that the famous Institute of Contemporary Art in London ran a chimpanzee art exhibit for three weeks in 1957. Artists from far around came to view the show, and some to ridicule it.[7]

Julian Huxley wrote in the New York Times, "These paintings are of considerable interest because they tell us something of the way art may have evolved…In fact we are witnessing the springs of art."[8] The show was a sell-out — famous artists purchased numerous paintings, among them Salvadore Dali who was impressed by Congo's clearly displayed ability to organize his paintings. Picasso also became an owner of one of Congo's works.

Interestingly, one of the many opinions relating to the origin of humans is that we are an "evolutionary sport," a sudden rapid mutation that stemmed from chimpanzees. Evolutionary sports are noted to have certain characteristics, such as a vastly reduced amount of body hair, and delayed maturity necessitating long-term care as infants. Such sports are often unable to become truly adult and remain youthful throughout life. Neil

Evernden, Associate Professor in the Faculty of Environmental Studies at York University, wrote in *The Natural Alien,* "The retention of juvenile form is called 'paedomorphosis' and the ability to breed while in juvenile form is called 'neoteny.'"[9] Chapter five of Evernden's book may be of interest to science teachers among others who seek an awareness of the dim glimmerings of our past, and who wonder about our inability to come to grips with many serious contemporary problems. Note incidentally the extent to which western people strive to remain young in appearance. Advertisements continually urge people to buy clothing designed to portray them as youthful. Heavy eaters are reminded that fatty tissue can be removed by surgery. Some individuals prefer this to dieting. Other surgical methods such as face lifts accompany changes in hair color, and heavy use of cosmetics depicts the apparent repugnance associated with ageing and our unwillingness to accept life for what it is — an interlude in which birth, growth, death, and decay follow one another as night follows day.

Inasmuch as organisms are dependent on the unique diversity of the ecosystems in which they live, the particular assemblage of animals, plants, soil conditions, temperatures and humidity of their indigenous locale best suit their development. Apparent minor changes in any of these can be severe or even lethal.

Consider some details in the life of Brazil nuts and note the specific relationships involved in their growth. This particular form of life is dependent on a particular species of bee to pollinate it. It is further dependent on a particular rodent, the agouti, which occupies the same forest habitat, and gnaws its hard shell and assures that some of the Brazil nuts are thus open and free to germinate to produce new trees. Furthermore the bee that pollinates the blossom of the Brazil nut is dependent for mating upon pollen from a particular species of orchid; and the necessary orchids need insects or hummingbirds to pollinate them in order that they may reproduce themselves. The foregoing intertwined symbioses are part of the diversity that enables Nature to maintain itself, and offer an example of how hasty human intervention in ecosystems can cause problems. The reduction of symbiotic relations is called simplification of the ecosystem.

Partnership with Nature

Simplifying the ecosystem by reducing the number of relationships actually reduces ecosystem stability. Responsible scientists have given warnings that we intervene in complex ecosystems long before we have knowledge of their intricacy. Actually the leading edge of the scientific community has for years stressed the idea that we will eventually have to admit that ecology merely recognizes limits in Nature that have always existed. The biotic potential (reproductive ability) of species simply runs into the wall of environmental resistance at the point where the reproductive rate outstrips the ability of the Earth to provide adequate sustenance. The noted scholar and scientific writer, Sir Julian Huxley, observed in the 1960s that, "The important sciences today for the modern world are not physics and chemistry and their application in technology, but evolutionary biology and ecology, and their applications in scientific conservation."[9] Huxley pointed out that a new ethical attitude must arise between man and Nature and that man can no longer falsely see himself as Lord of creation or conqueror of Nature. Instead he must become a partner with Nature. He called for an immense educational effort…excellence must be pursued rather than mediocrity, and diversity and variety rather than uniformity. We need to enhance perceptions, reflection and withdrawal. As Huxley noted, history indicates that the greatest of achievements in ethics, art, intellectual or scientific discovery occur only after a sufficient period of reflection. Instead of intense study of mechanistic topics, he observed, minds need to be allowed to lay fallow to help people escape semantic prisons. As Margaret Mead later suggested, universities need "Chairs of the Future," and perhaps also "(Huxley) Professorships of Possibility," or even a "Faculty of Human Possibility." Such faculties should be on a par with all other faculties.

By bringing an ecocentric view to the forefront in human minds we would be in a better position to cope with the severe problems that over-enthusiastic economic ambitions have brought to our finite planet. Certainly, if we can awaken from our lack of concern, a magnificent renewal could be provided in education, and students would be armed with the understanding to create the sustainable world our present mind-set excludes. The principal wealth we should seek is the abundance and variety of a diverse environment, and the power needed is the power to control ourselves and to become moderate consumers. Perhaps it is the immaturity of a sport species that prevents us from realizing that money is the Mephistophelean symbol that is successfully luring us to extinction. Billionaires quite likely are exponentially more distanced from reality than mere millionaires. There may be need for a certain amount of close budgeting and a sense of independence from the proverbial rat race to maintain full sanity. Instead of contenting ourselves with five-year plans for industrial growth we must act with respect for the future of generations yet to come. The Iroquois nation could recognize its responsibility to the seventh generation of its successors, but the corporate world, errantly pretending to be free enterprise, has its hand out for one subsidy after another — a failure from the beginning.

In every sense, we must recognize the Earth as our home place and thereby rise above our self-interest to understand that the protection of biodiversity is not a concession we grant out of the goodness of our hearts. It is rather a realization that if we do not protect biodiversity we will alternatively produce some resultant effect such as wiping out our own immune systems. We seem well on our way to doing that already. Our intelligence is obviously showing serious evidence of inadequacy.

As Ted Mosquin and Stan Rowe observe in their Manifesto, "Ecocentric ethics challenges today's economic globalization that ignores the ecological wisdom embedded in diverse cultures, and destroys them for short term profit." A peace-loving, bio-regional form of governance and studious efforts toward ecocentric behavior would move us closer to a sustainable society. As Ted Mosquin says, "By way of contrast, the present mode of globalization is at the root of the ecological and social justice catastrophe that is wiping out both evolutionary and cultural diversity. This is happening over ever larger regions and also destabilizing Earth's naturally evolved ecosystems, and most recently the climate of all the earth."

[1] Jerome Narby, *Intelligence in Nature* (New York: Penguin Group, 2005) p.159.

[2] Edward O. Wilson, *The Diversity of Life* (Cambridge, Mass.: The Belknap Press of Harvard Univ. Press, 1992) p.251.

[3] Ibid. p.251.

[4] Grace Herndon, *Cut and Run* (Telluride, Co.: Western Eye Press, 1991) p.139.

[5] Bill McKibben, "The End of Nature," *The New Yorker,* September 11, 1989.

[6] F.H. King, *Farmers of Forty Centuries* (Emmaus, Pennsylvania: Rodale Press, Inc., reprint from 1911) p.395.

[7] Desmond Morris, *Animal Days* (New York: William Morrow and Company, Inc., 1980) p.203.

[8] Ibid.

[9] Neil Evernden, *The Natural Alien* (Toronto/Buffalo/London: Univ. of Toronto Press, 1985) p. 14.

[10] Julian Huxley, *The Human Crisis* (Seattle: Univ. of Washington Press, 1963).

6: The Sixth Core Principle
Ecocentric Ethics Support Social Justice

Social justice for all humanity would be a natural outgrowth of justice toward Earth. But, before addressing this principle, let us also think of our relationships through time, of the likely possibility that regardless of race, nationality, or creed, we are related to one another. From our daily intake of food and drink we can realize our constant dependence on Earth. We are physically part of it, offspring of the same planet in the universe we inhabit. We also have uncountable relationships with many other life forms, soil types, and climate aspects. Note that there are both organic and inorganic relationships. In fact there are many necessities we rarely think of, that we classify as inorganic. These are of great importance.

Inorganic photons, for example, catalyze photosynthesis, the cellular process that underwrites plant and animal life. To grow and mature, living things also require inorganic soil minerals for health and strength. It has been a not uncommon hypothesis that all forms of life, plus their context, are fragments of an incomprehensible single Being, an immense Existence, which Stan Rowe sometimes referred to as a Supraorganism.

With that sort of a preface to relationships, I will steal a moment of your time to ponder the matter of numbers.

You may remember the thought-provoking question that used to be asked to illustrate the effect of continued doubling. Modernized, it would be, "Which would you choose: to work at, an eight hour job at a rate of ten dollars per hour; or at a rate of one cent per day with the wage rate doubling each day thereafter? At first thought many individuals would opt for the dependable ten dollars an hour rate. The more thoughtful individual will double numbers in his mind, 1, 2, 4, 8, 16, 32, 64, 128, 256, 512, 1024, 2048, 4096, 8192, 16384, 32,768, 65,536, etc. This reflective person will realize that although he will start work for one cent per day, the seventh day will pay 128 pennies or $1.28, and the sixteenth will pay 32,768 pennies or $327.68 for that single day. By the time the 34th doubling has taken place, the person will have received a number of pennies greater than the number of people living on Earth. The multiplier effect of doubling is indeed astounding.

The reason for giving the preceding example is to enhance the realization that the number of one's ancestors is enormous as we go back in time. What this adds up to in this prone-to-warfare world is that our individuality, as we call it, is itself made up of an incalculable multiplicity. Through descent we are related to billions of other souls, and even to the pre-human ones that have entered into the development of our own uniqueness. Our warfare becomes a form of treachery to at least some of our own ancestors, and is literally inter-familial warfare against our own relatives. This is not even a stretching of down-to-earth fact. Also, because of the terrible damage to Earth resulting from our barbarism, we inflict grievous wounds on the Earth, which is over-parent to all life in its myriad forms. We are thus rebellious Earth-wreckers in many of our activities. This, if you think about it, means that we should earnestly seek to live lightly. Since life entails choices, one can live quite pleasantly with no more negative impact than care can prevent. The old philosophical concept that life is a continuing choice between the sacred and the profane may come to mind on this issue. Thinking about this idea might actually help an individual to make wiser choices by asking oneself which of any two choices is closer to the sacred than the other. Apologies — I am not trying to be profound, but only to indicate the continuing judgment that life entails.

Our relatedness to incalculable millions of other beings, some of shapes and characteristics we cannot begin to conceive, should put an end to the sinister and foolish worldview initiated by industrialism that humans are superior to Nature but not actually part of the natural world. We have been taught to turn up our noses at the idea of membership in the animal kingdom. Industrialism's determined supporters still argue that only humans have souls, consciousness, and intelligence, whereas other beings merely function by "blind instinct." Such spurious reasoning has affected easily influenced, opportunistic individuals, and amounts to a deliberate de-sanctification of Nature, which was for generations considered to be holy. Industrialists, adopting this illogical but handy attitude, have ruthlessly devastated ecosystems at will though traditional thought held that this would invite the wrath of the gods. Allegorically, it has done just that.

Corporations Flawed

The Manifesto says that, "Many of the injustices within human society hinge on inequality. As such they comprise a subset of the larger injustices and inequities visited by humans on Earth's ecosystems and their species. With its extended forms of community, ecocentrism emphasizes the importance of all interactive components of Earth, including many whose functions are largely unknown."

Now, in our society, industrialism has virtually been given *carte blanche* to do as it pleases. Instead of having government closely and continually supervising corporations, they do as they wish and government looks subserviently the other way. Here are some of the injustices committed against ecosystems, all resulting in social injustices to all of humanity as well by virtue of our relationship to the ecosphere.

The Caterpillar Company specially modified D9 and D10 bulldozers for the Israeli government to facilitate the destruction of Palestinian houses. Caterpillar has refused to end its corporate participation in house demolition and continues to provide these machines to the Israeli military forces. A number of Palestinian people have been killed in their homes by these bulldozers. Here we have irresponsible behavior on the part of a corporation and on the part of the Israeli government. Notably we see here collusion between corporatism and governance in the convenient destruction of homes, lives, and the personal possessions of many people. Social justice is mocked by irresponsible modern power structures.[1]

It is not surprising that Dow Chemical has been identified as socially bankrupt. It has been named as a business entity that "has been destroying lives and poisoning the planet for decades." Its development of Agent Orange led to health disasters for millions of Vietnamese citizens and many US veterans of the Vietnam War. Dow also developed napalm, a furiously burning chemical weapon that killed many people in Vietnam and other wars. In Plymouth, New Zealand where 500,000 gallons of Agent Orange were manufactured, "thousands of tons of dioxin-laced wastes were dumped in agricultural fields." Both government and Dow Chemical have displayed stunning sadistic behavior. Dow displayed ruthless cynicism by developing and producing these poisons, and government revealed its unworthiness by abetting production and then using such barbaric weaponry. No concept of social justice is evident on the part of either of the instigating parties.

Nestle USA has been cited for illegal and forced child labor. The US State Department estimated that 109,000 child laborers worked in hazardous conditions on cocoa farms on the Ivory Coast. In 2000, Save the Children Canada reported that 15,000 children between nine and twelve years old, many from Mali, were tricked and sold into slavery on West African cocoa farms. The sale price for each of these children was often about thirty dollars.[2]

Monsanto produces many pesticides sold to farmers, including Roundup (glyphosate). Its toxins accumulate in soil, rendering the soil infertile. Farmers are then encouraged to purchase Roundup Ready seed, which is resistant to the pesticide. This enables Monsanto to create dependency for its pesticide and its resistant seed. Both of these are sold for inflated prices. Exposure to Roundup can cause cancer, skin maladies, spontaneous abortions, premature births, plus gastrointestinal and nervous system disorders. These products are an excellent example of the hidden threat of high (?) technology.

According to reliable sources, Monsanto also employs child labor. In India an estimated 12,375 children are utilized in cottonseed production for farmers hired by Indian and multinational seed companies, among which Monsanto is included.[3]

It would be simple, although tedious and repetitious, to name and elucidate the long list of socially irresponsible actions on the part of other corporations and businesses. Once again we are seeing examples of chauvinistic homocentrism. A 2006 study from St. Andrews University, Scotland shows that fewer than 4 percent of the world's 50,000 major corporations produce reports on corporate social responsibility. Assurances that corporations have assessed their social responsibility are "at best useless and at worst highly misleading." The authors of the study believe that sustainability will be emasculated by corporations. An author of the report says, "We believe we must treat the current crop of 'sustainability reports' with the most profound mistrust, as one of the most dangerous trends working against any possibility of a sustainable future…Unless we change the way the world is organized, we risk even greater social injustice and more ecological disasters…Driven by globalization, problems of pollution, waste and global warming are all threatening to disrupt humanity in unprecedented ways. Controlling the multinational corporations that cause some of these problems is not going to be easy."[4]

A thought pops into mind that has to do with the establishment of corporations, and particularly with the legal anomaly that deems a corporation to be an artificial person, nonetheless with rights of a person in law. Something smacks of hubris in this irrational legality. The creation of these artificial persons may explain why many corporate deeds show the absence of a conscience in their performance. This seems to be a travesty of law that calls for righting.

The dominance of humanism, a belief that man is the measure of all things, has led to other "isms." These include an altered form of individualism, which has approved selfishness, disregard of responsibility to others, and refusal to accept constraints that might lead to a more stable society; materialism, which has to do with surrounding oneself with possessions and which Edward Goldsmith suggested is the true opiate of the people; scientism, the narrow view that our limited, and sometimes superficial scientific knowledge, should be the dominant factor in both social and ecological control; technologism, the obviously fallacious belief that technology can cure any problem, which is now refuted by immense evidence that aberrant technology is destroying the planet; institutionalism, the view that institutions should govern Nature, which is childish in that Nature's own self-regulating systems are warning us with vicious storms, unpredictable climatic effects, and increasing tectonic activity, that our irresponsible behavior has been overruled; and economism, which suggests that things should be done in such manner as to provide the greatest possible return on capital.

The one appropriate "ism" would be "ecologism," which is the understanding that every act performed affects the integrity of the ecosphere. These acts lead to complex or simple responses that are injurious or beneficial in their performance. A species in harmony with planetary integrity can survive, while one that violates planetary integrity is eventually discarded.[5] Therefore, in regard to all human activities, impact on the ecology of the Earth should be of vital concern and should not be ignored, as it frequently is for homocentric purposes.

The Great School of Life

The bodies in which we live our transient lives are impermanent agglomerations of individual units called cells, which exist in astronomical numbers in each human body. What is the unique human body/mind but the assemblage of quintillions, sextillions, or numberless life experiences of those beings that have preceded us? Is there really any entirely new thought or emotion; or do we relive the life experiences and concepts that have been repeated, though with infinitely tiny variation, in each of us who has emerged from the dust, perhaps to gain some microcosmically greater insight into life? Is there any greater school than life itself; and should our own schools attempt any more than to prepare us for greater understanding of life and possible meanings of this quest? Humanity is like the tip of an iceberg rooted in the vestiges of past lives. Life apparently dissolves and reassembles, but the time is always now, and present forms of life rest on shadow forms that have preceded them. The continuity of our past may be why some people seem to recall things they cannot explain. Is a warning reflection merely a prompting from a shadow past? Are there any brand new thoughts or original feelings? Are we anything other than bit parts of a stupendous whole and is a soul anything other than a fragment of the One Soul? Many people concur that there must be some meaning to life, some purpose to be sought, some insight to be gained. Perhaps we are here to serve the wholeness of Earth. As Longfellow suggested:

> Tell me not, in mournful numbers,
> Life is but an empty dream,

For the soul is dead that slumbers,
And things are not what they seem.
Life is real! Life is earnest!
And the grave is not its goal;
Dust thou art, to dust returneth,
Was not spoken of the soul.

I refer to life as a school because I believe that life is a learning experience. It also seems to me that people learn more about social injustice than about what social justice should be. Is life trying to tell us something? I think fondly of a long-gone friend of mine, Patrick James Carroll. He and I walked, talked, prospected, and philosophized together. Parts of his life had been lived in turbulent depression years unlike the affluent years of today. Tough economic conditions in the early twentieth century, which put him on the road as a hobo, brought unforgettable experiences that shaped a strong character.

Paddy, as he was called, had a very large dose of experiences cast in his lap. He adapted and survived. For years he slept where life found him. He learned that one could sleep in "crumb dumps" as they were known. For a nickel you could buy a large newspaper for your mattress and sleep on a floor under cover from rain and snow. On a Christmas evening he was attracted to mass in a Catholic church along his way. At that time, he said, there was an offering stand by the doorway where people customarily donated fifteen cents. He said that his quest for formal religion vaporized when he was turned away because he had no money and wore bedraggled clothing. He ironically commented that the law in all its majesty and equality forbids both the rich and the poor to sleep on park benches.

He told me that as fall approached, "We bums tended to move south, where the climate suited our clothes." When he got into balmier climates, he and numerous others sharing the same state of affluence were stopped by sheriff's deputies and asked to show their money. If they had none, or only a pittance, they were impressed into working on the levees for their keep. In the spring they paid us off, he said, by chasing us into the swamps with a pickaxe handle. He completed his tale, musingly saying, "And that was in the Land of Liberty."

He headed north into the Ootsa country in British Columbia in 1915 and became a packer with the famous Cataline and his 60-mule pack train.[6] Later he and his brother started their own pack string, which carried goods from Burns Lake to Smithers. Most stories such as Paddy's do not end on an upbeat note. But, as he said, he always looked at the ground and studied the minerals as he walked. Thus as he approached seventy years of age, he discovered what became the Granisle copper mine, the largest low-grade copper property of that time. He became very wealthy and equally unconcerned with wealth. He also became a poet, a sort of Robert Service who actually lived the life about which he wrote. That he found social justice in Nature is pretty well conveyed in these lines:

Harnessed to the car of commerce
I shall e'er refuse to be,
I have found a habitation
In the mountains, fresh and free;
Far from strife and din of cities
I have found a place of rest;
Here amidst the snow-capped mountains
In the valleys of the West.
Purple hills and snow-capped ranges,
Emerald lakes where trout arise,
Is to me a place of refuge
'Neath the sunny northern skies.
Grouse are drumming in the forest,
Moose are roaming on the hills;
Rugged, ragged Omineca,
How my heart with rapture thrills.

Paradox seems inherent in life. The sheriff's men who chased bums into the swamp would have been surprised to know that one of the bums, Paddy, was a direct descendant of Charles Carroll of Carrolton, whose name appeared as one of the signers of the American Declaration of Independence. Strange indeed, but true.

Ethics Stem from Maturity

This principle of the Manifesto is about the fact that our ethics should be derived from all the interactive components of the planet, and with respect for the assemblage of symbioses, feedback loops, and entwinements that meld all life into the spectacular ecosphere. The Manifesto views the injustices within society as a product of the hierarchical structure of an arrogant species that discriminates particularly against disadvantaged women and children, and against the powerless and the poor. It views with alarm the rapid degradation of ecosystems, and the increase in human tensions as inequality increases and selfishness is justified, while suffering is ignored. It stresses the critical need for altering our worldview from homocentrism to ecocentrism. Literally though, an ecocentric ethic can only exist when a person has studied enough about Earth and the life on it to comprehend how our relationships with Earth and with one another should be formulated.

Seneca's expressiveness provides a good analogy that might be kept in mind, namely, "Our relations with each other are like a stone arch, which would collapse if the stones did not mutually support each other, and which is held up in this very way."[7]

Pondering the foregoing paragraph should suggest to you the origin of many of the problems that threaten society. It is a well-known fact that governments and agencies of all sorts have long used the technique of divide and conquer to rend society into malleable factions. Throughout society many people continually foster competitiveness, and unreflectively throttle cooperation. In short, towns, cities and nations are set against one another in many ways. Stressing the I'm-better-than-you attitude, we stifle the idea that people are all in the same boat, all of us are part of life, and all of us are striving to keep our heads above water. This belief is brought to the level of governance where governments are formed of parties that compete with one another and exaggerate their differences in a continuous no-win battle. Though differences may be miniscule, these differences become fodder for the oratorical emptiness of political speeches, which are so elusive that we must quiz ourselves to determine what the speakers really mean. Pounding on discordant political drums has enormously exaggerated the differences between races, nationalities, and creeds until bickering and warfare have become as inevitable as the common cold. Looters, looking for every opportunity for profit, continually turn out new weapons to assure that the cost of imminent war must be borne by every nation that fears to be unprepared for the next round of chaos. Our species is trotting along on the extinction trail, refusing to realize that people are part of the ecosphere and cannot afford war.

So, there is one thing that people share and, in spite of our desire to be the focus of the universe, we must face the simpler truth that the Earth is the common denominator for all beings. Since it is a oneness, we cannot damage it in one place without damaging the whole. It is axiomatic that if we display justice to Earth we will as a matter of necessity have to extend it to one another.

The essence of the whole matter is that a sense of social justice must be derived from our own thoughts and understandings. Social justice is unlike a 10 percent tithe. It is a conviction striving to be born from necessity. It requires an understanding of infinity such as that expressed by the second law of thermodynamics, that neither matter nor energy can be created or destroyed, although they can turn from one form to the other. Future generations are already slated to suffer from our ubiquitous poisonous impact on the ecosphere. Glutted by materialistic habits, western society must call upon spiritual ideals to practice self-denial and apply the initiative necessary to restore the world before it is too late, if it has not already reached that point.

Let's peruse the matter of modern warfare a bit farther. An article from Science Action Coalition reports, "Man's Great War...the war against Earth and life continues unabated. During the Vietnam War the US Defense Department delivered an average of thirty tons of explosives per square mile in Vietnam. Just for the sake of comparison, from 1965 to 1975 strip-mining operations in Appalachian coal fields detonated thirty-five tons of explosives per square mile." Meanwhile, a UNEP report states that in South Vietnam, chemical herbicides "completely destroyed 1,500 sq. kilometers of mangrove forest, and caused some damage to about 15,000 sq. kilometers more."

The foregoing information does indicate the tremendous "business revenue" that fortuitously drops into the pockets of manufacturers of munitions whenever a war takes place. Furthermore, it is troubling that Canadian pension plan investments in the top ten US military contractors totaled nearly $200 million on September 30, 2003 and $650 million by March 31, 2006.[8] I find it ethically repugnant that Canada should support world militarism in this way while describing itself as a peaceful nation. I must admit that I am troubled by Canada's participation in the Afghanistan debacle. The possibility that military ventures are sufficiently profitable to cause this nation to become an investor in the manufacturer of munitions, and also a participant in another war upon the Earth (for all warfare is ultimately planetary degradation) appears to me as a revelation of the ecological blindness possessed by governance. The already existing ecological degradation in Afghanistan is sufficient evidence of human nature that is deeply in need of massive change.

A Bit of History

Consider what is currently taking place in Afghanistan, and how attempting to provide social justice to the people might be an alternative to warfare. An extreme climate and inadequate rainfall have combined to make Afghanistan a land to test the strength of its people. It is believed that in early times the harsh winters led to indiscriminate cutting of forests in order that the people might have fuel.

Prior to 1220 AD, the Mongols invaded Afghanistan. Historian, Sir Percy Sykes, gives us some idea of the extent of the damage that they inflicted: "Bamian had been captured…and to avenge the death of one of his grandsons, Chenghiz (Genghis) destroyed every living creature, including animals and plants, and the site remained desolate for a century."[9] History bears out the fact that Attila expressed the rapacity of many military conquerors, including today's, when he boasted that, "no grass grew where his horses' hooves had trod."

At Herat, a city destroyed by Genghis Khan in 1226 AD, the destruction was so merciless and thorough that it is still visible in the twenty-first century. The entire population of Herat was massacred. Desert sands blew in over the plowed fields. The kanats collapsed, and to this day there are kanats that have not been repaired since the Mongol invasion. Without this life-giving irrigation system the entire agricultural prospect of Afghanistan rapidly declined. In the 1930s an irrigation program was begun in the valley of the Helmand River, but failed to take into account a people unused to adequate water supplies. Thousands of acres unfit for irrigation were over-irrigated. This caused water tables to rise and the soil to become salinized. Expensive reclamation projects thus had to be undertaken to make the soils arable once again. It becomes evident that damage to the land, while perhaps aimed at only a single conquest, can lead to a poorer life for untold generations of people who will live long after the folly of the military victories has become apparent.[10]

As the Rienows stated in *Moment in the Sun,* the main problem in Afghanistan is its "irremediable pollution…not only are the surface waters laden with diseases and plagues, but the lack of proper sewage disposal over the centuries has caused permeation by pollution of the sparse soils until the deep aquifer — that vast and mythical lake of fresh water underlying Earth's surface — has become universally contaminated, and there is excessive contamination everywhere in the land."[11]

The foregoing is a vivid example of the impact of warfare. The current military action in Afghanistan will merely add to the long-term damage to Earth. The ecocentric solution would be to cancel the war and commence the lengthy process to restore the land, which would truly trickle down to restore its habitability. It would be a long and difficult task and would tax human skill to the utmost. However, if we cannot accomplish that sort of task, how can we possibly justify the rather childish aspirations some people have to become spacemen? If we cannot heal the inner space of our ecosphere and our unremitting egotism then why pretend that by blasting holes through our atmosphere and ozone layer we can conquer outer space?

It is apparent that people should cease military warfare and commence a form of restorative warfare against the atrocities our species has conducted against our home planet. I suggest that this would mark the maturation of our species, and the beginning of a realistic approach to world problems.

Call to Femininity

Specifically I believe that this is a greater call to femininity than has taken place at any time in history. Women have traditionally been the nurturers of all forms of life. They have swept to their bosoms not alone the care of children, but the care of puppies, kittens, injured or hungry birds, of all forms of life, even tending the wee turtles, tiny frogs, gerbils and stray creatures of all sorts brought to them. They have been our teachers, our nurses, and the stability of families. They also seem to display more genuine concern for what is being done to the Earth. Now the call is upon them from the Earth itself to bring up the older generation of age-matured but toy-centered males still fascinated with things that spin their wheels, snarl along roads, or tear through waterways with a roar and a cascade of water, foam, and wave. This includes mankind's even greater fascination with diving planes, exploding bombs, torpedoes, and even with the shivery delight of contemplating cataclysmic nuclear war.

It is aging Mother Earth that has been stricken by ruthless destruction of waterways, forests, seas, atmosphere, and even farmland now regularly saturated with poisons for the supposed nourishment of all families at their tables. Mothers have been the traditional protectors of society, and their sense of conviction has always had a ring of reality about it that has strengthened society. It is mostly men who have made the Faustian bargain to sell their souls for technologies and power, both of which are clearly out of human control. It is remarkable that Goethe had the insight long ago to recognize that alienation from Nature is the greatest mistake to which a person or even a species may succumb.

Our world of materialism continues because many people curb their thoughts so that they cannot venture into any concepts of greater meaning. Like Aesop's fly, perched on the axle tree of a chariot rolling down a dusty road, they focus only enough to say, "My, just look at the dust I am raising!" It raises to a species level the observation that "a man in love with himself will have no rivals."

An Unusual Experience

For four years I wrote a three-to-four-minute, weekly radio script for CBC in Calgary. In nice weather I would go to a nearby creek and sit on a large cedar log that spanned the stream's twenty-foot width. I enjoyed the serenity of watching the water flowing downstream and was pleased to watch the antics of occasional water ouzels that I saw. I rarely thought about the script I would later write that day but did feel some sort of spiritual exhilaration absorbed from the pristine setting around me. On the special day that I remember, my reverie was interrupted by a slight sensation at the back of my neck, which became an acute prickling. Instinct turned my head to the left — I was facing downstream — and I felt a surge, of adrenalin I suppose, for there was a medium-size grizzly bear looking at me, one forefoot lifted atop the log. I said nothing. Anything I said would have been superfluous and perhaps unwise. When one is sitting, with feet dangling above the water, it is difficult to get up gracefully, and I admit to feeling awkward. But I did stand, and saw that the bear had not moved. I turned my back to the bear and walked atop the log until I could step down onto the ground, and then continued walking downstream, turning my head just enough to note after I had taken a half dozen steps, that the grizzly crossed the stream to my side and, in what I think of as a very polite manner, ambled upstream. I had heard of the erector pili muscle that causes hair to stand on end, and later realized that some ancient bodily memory somehow became activated.

I have often reviewed this incident, as I do now, and have pondered which of my five senses was altered. I also wonder about that composite sixth sense we perhaps all have. Racial memory? Who knows? In reflection I can say that the experience was one I could not account for in any way. It certainly had an element in it that was frightening, which was negated considerably by the demeanor and calmness of the bear. In memory, I also feel a sense of privilege. I have great respect and liking for the wild creatures with which I like to coexist, and believe that an individual on a city street, perhaps now at almost any time of day, is in far greater danger than someone walking through or sitting in the relative wilderness we have today. More than two-score years living in the woods makes me more inclined to remain there, and less inclined to visit "Urbania." My mental synopsis of the whole matter is that I was accorded a priceless experience for which I feel great appreciation. To the cosmos? Why not?

My friend Alan Marlow told me of an even closer encounter one day last summer, when he was preoccupied with a small piece of machinery he was studying as he sat on his steps. An animal was moving around within two or three feet of him. In his concentration he assumed it was the neighbor's dog, and did not look up. When it sniffed his foot he glanced up and saw that it was a black bear. Not particularly perturbed, he told it to "get out of here," and it immediately ran off — no doubt frightened by this stationary object that had suddenly come to life. Alan, who moved into this country in 1929 as a small boy, remarked that he and his brother and two sisters, grew up here when it was an almost daily occurrence to see both black and grizzly bears, and he said that the bears were never a threat. "Good neighbors," he says.

Life is Grace

I sometimes wonder whether we live in periodic but perpetual palingenesis, or reincarnation, as it is more popularly known. I also wonder whether I am most comfortable living in a woodland setting because it is where I feel the greatest serenity. Linda is also comfortable here in the same manner, and often when we return from a trip to town she expresses how good it is to be home. I have felt since boyhood that the forest country is where I naturally belong. So I do wonder if some inborn attraction to Nature is part of my own heritage. The old myths of bodily resurrection may be foreshadowing of truths that far surpass the fables and religions we are driven to follow and accept. What is life but a privilege and a form of grace we are too immature to appreciate? And is the intent of the universe anything other than that we must, however we protest, become worthy of the grace we have received for millennia? Are we anything but fatuous to deny higher levels of intent than our own? Does this not mean that we must transcend our greed, our foppishness, our devotion to constant mobility, and dilettante tourism? Must we, can we, arrive at the political level where idealistic guardians of the public good replace selfish demagogues focused on power and an imaginary niche in some unread history book?

Ecocentrism is itself the root of a sustainable philosophy that can foster a civilization in which social justice would embrace all the interactive components of Earth, living and non-living. Our foolish focus on conspicuous consumption would be replaced by a wiser concentration on ecocentric necessities. Activities should get underway that alleviate some of the serious ecological problems we have caused, thus providing a mindset capable of resolving issues of social justice.

Where Does Social Justice Begin?

Social justice must *commence* with the sort of respect and love that is accorded family members, and these feelings must extend to the cosmos "in which we live and move and have our being." It interests me that this phrase, as found in the New Testament, was actually derived from the Stoics who felt that Nature is God. Seneca, one of the foremost of the Stoics, expressed the thought in this way: "All that you see, all that encompasses both God and man, is one — we are all parts of one great body."[12] Seneca's thought seems sensible and is an early example of ecocentric thought.

Problems of social injustice cannot be solved by those in power making empty promises while steadily allowing degradation of the planet. They are seemingly unaware that the integrity of Planet Earth is the dynamic wealth that makes Earth livable. That integrity accounts for our survival.

Our leaders are wrapped in trivial pursuits that, in true diabolic fashion, blind them to what they are doing. They have dulled the sensitivity that would enable them to understand they have been violating ecospheric limits that will not idly accept endless exploitation. The barrage of empty words and irrelevant deeds continues from governments that steadily procrastinate while driving the planet and the life it supports toward an irreversible collapse. Industries maintain their myopic and psychotic fascination with conquering Nature while destroying its integrity. Economic interests are proving that "the love of money is the root of all evil."

Systematic methods used to divert human attention from our natural heritage have been so successful that we are approaching uniform collapse of climate, economy, and civil society. We have become indifferent clones and have been led to believe that we must keep up with a world going nowhere. A few days ago, a visitor who works for a telephone company told me he has noted, with a degree of contempt, an escalation of what his company expects from employees. For a number of years his employer has called for 110 percent of an employee's effort. He noted that the number continues to increase, and was told a short while ago that he should give 140 percent of his effort to his job. This may seem trivial but is not surprising in a world wherein people are expected to act as sterile machines. Slavery in America has been replaced by the expectation that people should become robots for corporations.

Taking Responsibility

It seems that we are well on the way to extinction or to a vast reduction in population. Perhaps the reason is that we have given up thinking for ourselves and surrendered our consciences to ludicrous leaders. We should react by driving less, flying only when absolutely necessary, staying home and taking walking holidays rather than serving the tourist industry. We can have gardens in whatever space we can innovate and should develop the habit of buying only what we need. Note the frequency with which we are supposed to spend lavishly on one of the many occasions dreamed up by advertisers. Take Christmas as an example, and note that the day celebrates the birth of an impoverished child born in a manger, an individual who stood for strength of spirit and devotion to principles but not for ostentatious display.

We must regard with suspicion, advice to spend without discretion.

We should accept responsibility for our own security and look askance at advertisers who encourage us to mortgage our homes for all we can get and then spend the proceeds lavishly for whatever we want. There are many regulations to curb individual behavior but very few, if any, to curb the pouring forth of suggestions from various forms of media that we spend perilously and thus jeopardize our futures. Governments have succumbed injudiciously to business interests and accord such interests the liberty to perform constant social injustices. In this day of sophisticated (falsely wise) blandishments to consume recklessly, there should be countervailing advice from government that advises people to be cautious consumers, and safeguard their homes and families from gangsters in the guise of advisors. What I am saying here is that there should be as great a separation between government and business-gone-wild as there has traditionally been between church and state. Government — at least since the time of President Coolidge in the United States, and at least since the time of Prime Minister Mulroney in Canada — has allowed business to be in the driver's seat and has seriously aborted its responsibility toward the people en masse. Coolidge's policy was expressed as "more business in government and less government in business," while Mulroney's dictum showed disregard for the ordinary citizen, as expressed in his words, "The business of Canada is business." Consider the spiritual atrophy evidenced in a statement made by a federal government minister of Canada in the 1980s: "There's one underlying motive in business shared by all — it's greed. We support it whenever it happens."[13] How Nemesis must smile to hear such words! And in reality, do Canadians wish to have their government act with benevolence toward the champions of greed while many less fortunate citizens struggle to make ends meet on minimum wage salaries? It seems as though the public is endlessly hoping for governments to awaken.

To gain a better idea of the plight of citizens in the US, we listened this morning to an interview on CBC radio with Lewis Lapham, editor emeritus of *Harper's Magazine.* Lapham's view, shared by others, was that George Bush has been guilty of criminal behavior and should be impeached. Several other people were interviewed regarding Lapham's opinions and one of them pointed out that if Bush was removed from office he would be succeeded by Dick Cheney, who he suggested would not be any better. This is the sad state of affairs that has resulted from the melding of government and business as mutually selfish entities. We should have grown up a bit more than the characters in George Orwell's

Animal Farm who echoed today's blatant injustice in the ideological statement, "All pigs are equal, but some are more equal than others."

In contemporary society, the "more equal pigs," which refers here to big business, have succeeded in violating the stability of communities, in reducing the health of land for immediate profit, and being injudicious and uncaring enough to do harm to community stability and to total food supply. For example:

At St. Peter's Abbey in Saskatchewan, Roman Catholic Bishops of Saskatchewan gathered in 1987 to study agriculture and issued their conclusions in the document: "Farming — A Vanishing Way of Life." Considering the economic, social and ecological conditions of agriculture in Canada they reached the conclusion that "present agricultural practices are not in accord with the Christian concept of stewardship." The reason for this they decided was because the health of the land and the well being of citizens were subordinated to profit and what is called "efficiency" today. Their statement conveyed in Ron Graham's book, *God's Dominion — A Skeptic's Quest* was this: "The inadequacy of our current system is vividly reflected in the death of small villages, in rural poverty, in the lack of young people interested in farming, in dwindling rural populations, and in the increase in hunger and malnutrition both in the third world and here in Saskatchewan."[14]

Our society has deemed that everything else is secondary to the pursuit of money, this pursuit being exemplified in what is called business. Considering the impact on agriculture of small communities and on farming as the promotion of good husbandry and care for the health of land, we have cast both social justice and good ecology to the state of non-entities. Small farms, far more productive than large ones, have been virtually wiped out by the metastasis of bigger businesses. Thousands of good farmers, producing genuine food, utilizing crop rotation and farm-produced organic fertilizer have been excluded from their way of life by greed, vastly encouraged by governments.

From where I live, an individual must take a ferry across Arrow Lake and then drive about thirty miles (55 km) to visit Revelstoke. Before Mica dam was built there were many small, productive farms on both sides of the Columbia River. These were sacrificed for power and ignominiously drowned out along with the hopes of their former owners. The landscape was lovely and one was treated to the sight of pastured livestock, orchards, of farmers and their families engaged in haying, cultivating and other productive activities. People were leading meaningful lives on private properties. Also, a large amount of food was produced for local use, which means that some of the food that reaches our tables must now travel 2,000 miles. If we thought about it seriously, we would realize that localization of food production is much safer than dependence on a globalized food supply.

Human desire for endless and poorly managed power supplies led to writing off many homes and much valuable farmland for power used for many unnecessary reasons. Social justice was ignored and suffering was severe and widespread. I remember going to Revelstoke one day after the farms had been taken from their owners. I was alone and sat at a counter eating lunch and conversed with the man sitting next to me. It turned out that he was one of the dispossessed farmers who had lost his way of life and the home he had owned for more than thirty years. His bitterness was real and sadly tragic. He foresaw that the day would come when humans would finally realize that decency and stability are regularly destroyed by an unrealistic concept of progress. He understood intuitively the negative aspects of losing agricultural land, forests, and wildlife to flooding. He understood also that the fishery in the Arrow Lakes would no longer be nourished by silt carried in the river. This silt would now be trapped behind the dam. His vision was correct, for now it is necessary for the Arrow Lakes to be fertilized by thousands of gallons of chemical fertilizer each year. Elsewhere in the world, dam building has increased the incidence of malaria, schistosomiasis, filariasis, onchocerciasis (river-blindness) as well as causing other problems that are very serious — salinization being one of them. These problems have become worldwide, and man is their root cause.

A massive program of de-development could bring about much employment and the improvement of Earth health.

Overemphasis on the profit motive is somehow on a collision course with traditional attitudes of social justice. It seems a gross simplification of human responsibility for a single and extremely parasitical segment of society to interfere with the very foundations of what, not too long ago, was a reasonably stable society. All forms of zealous activity can progress to serious sociopathic levels, and that is where the promotion of business as the zenith of human endeavor has led. Perhaps the next G8 summit get-together should consider the ancient adage, "What profits it a man to gain the whole world and lose his own soul?" They might also discuss the philosophical observation made by Henry Thoreau that, "A man is rich in proportion to the number of things he can get along without."

The Need for Personal Efforts

There is no question that some of the extra money individuals have can be used in ways that would benefit the ecosystems in which they live. Years ago Linda and I bought a small acreage that had been largely clear-cut and was left littered with waste. We bought it to reforest it, not for future profit, but because we recognized it as a thing that should be done. We happen to believe literally that Earth is the "set" and we humans are one of many minor subsets. I would also say that we recognize and respect the ancient thought that sincere human beings should, as a matter of conviction, willingly and actively seek "to dress and keep the Earth." We even realize that it is a privilege to try to restore the Earth. I say this knowing that the Earth doesn't need us at all and that one of our real purposes in life would be to avoid harming the Earth, because in the final analysis we are of the same stuff as is Earth. Excess wealth that people have extorted from the planet and one another could be spent in canny and useful ways. Prominent human beings seem seriously concerned about being remembered in history books and in memorials. As for the quest of many individuals to have things named after them, they could go out and plant trees or repair a human-made mess without the need for applause. If they put up a sign to say what they had done, it would fall down in time, but the things they planted or land they had improved would continue to affect eternity in some positive way.

Wealthy individuals could establish memorials that would truly honor them by virtue of the real usefulness their memorials might perform. I would like to suggest that Joe, Jane or Jim Billionaire could perform a very valuable service to the world by leaving behind (not a pun) a world-class sewage plant named in his or her memory, meanwhile serving the region in which he or she lives. Certainly it is time humans faced the problem of pollution created by their own excrement and the resulting problem of polluted waterways and oceans, which are the ultimate recipients of untold amounts of wastes, toxic and otherwise. First we would have to conquer our irrational delicacy, and reluctance to pay any attention to serious problems that might suggest we are part of the animal kingdom. Sewage recycling is such an enormous issue that it should be included in school curricula.

Let me illustrate that this problem has been seriously discussed in classical literature. I give this example because the text is available in both French and English. While we deign to face this problem, which has vital possibilities for helping to restore the planet, it would be a good project for high school students to take on as outside reading, Victor Hugo's classic, *Les Miserables*. It would also be excellent instruction for the chapter entitled "The Intestines of Leviathan," to be read aloud in many classes as an example of historical observation of the unwise behavior of a species such as ours, which dumps valuable human manure into a nation's major waterways.

Victor Hugo starts off the chapter by speaking of the fact that Paris throws vast wealth into the sea by means of the city's sewer, which he refers to as its intestine. He reveals in this sense a major economic problem of the world that has been continually ignored, and that involves "the most fertilizing and effective of manures." He states quite accurately that, "A great city is the most powerful of dung producers," and emphasizes the wealth represented by the sewage when he writes, "our manure is gold."[15] He describes the manure of humanity as "the fetid streams of fetid slime that the pavement hides from you," and then asks us to realize what the wealth of that resource really represents.

In answering his own question he clarifies once again the miracle we disdain, which is the magical, miraculous ability of Earth's own processes to transform fetid putridity into beauty, health, and food. This virtual resurrection of waste to vibrant life is expressed in Victor Hugo's recognition of what the waste becomes: "It is the flowering meadow, it is the green grass, it is marjoram and thyme and sage, it is game, it is cattle, it is the satisfied lowing of huge oxen in the evening, it is perfumed hay, it is golden wheat, it is bread on your table, it is warm blood in your veins, it is health, it is joy, it is life. So wills that mysterious creation, transformation on Earth and transfiguration in heaven."

Victor Hugo does go on to satirize mildly the "cleverness" of a species that throws hundreds of millions of francs, dollars, guilders, krones, or whatever kind of money is named, into the rivers and thence into the oceans, denying the restoration of Earth energies that have been bestowed on all life. He points out that our whole economic system, in its ignorance, throws our wealth into the gutter, eventually causing hunger to rise from artificially fertilized agriculture, and disease to arise from the rivers that serve us for drinking, for washing our bodies and clothing, and for other essential purposes. Homo sapiens is obviously not as sapient as the species pretends.

Another French writer, Emile Zola, in his novel *The Earth,* states that the dung of Paris is adequate to fertilize 75,000 acres, saying, "The great city would be restoring to the land the life which it has received from it. The soil would soak up these riches and the fertile, bloated land, would lavish harvests of good wheaten bread."

Pertinent to this discussion, Zola also spoke of education, saying, "Make a point of it, our salvation can perhaps come from proper schooling, if there's still time!"

Further to the foregoing, Victor Hugo wrote in his manuscript *En Voyage, Alpes and Pyrenees* (1891–1894): "In the relation with the animals, with the flowers, with the object of creation, there is a whole ethic scarcely seen as yet, but which will eventually break through into the light and be the corollary and the complement to human ethics…It is also necessary to civilize humans in relation to Nature. There everything remains to be done." Social justice issues cannot be resolved until the rapid degradation of Earth's resources is halted. Only then will there commence to be hope for our species to survive in a civilized state. You might also notice that Hugo's words are a fine example of ecocentric focus.

Ecocentric Ethic Fundamental to Social Justice

It seems valid to suggest that before we can really achieve social justice, we must recognize that our surrogate society is far out on the tip of a limb that it is sawing off with frantic haste. There is a deep wisdom in recognizing "diversity with equality" as the mature viewpoint that would enable us to begin the job of reapplying for full-scale membership in Earth's family of living things. It is difficult to determine how and why our species went astray. It could be that society's organizing ability is both its strength, and its weakness. How is that so? Consider how we discriminate in our own favor. For example, people, wherever they may be, have more than a tendency to mock the religion of Native people as pagan. They believe in a Great Spirit that permeates all of Nature. We ridicule that concept, and yet admire the same idea when wrapped in the fancier beliefs of modern day religions that deity is omnipresent.

In a book entitled *Philosophy — The Basic Issues,* there is an essay written by H.L. Mencken titled, "Memorial Service."[16] The article starts out by asking, "Where is the graveyard of dead gods?" He asks, for example, what has happened to Huitzilopochtli, a forgotten god to whom 50,000 youth and maidens were slain in sacrifice? Like many other gods, he was considered omnipotent. Like many others, his appearance was miraculous. Like more than 120 other gods named in Mencken's essay, he was immortal, a god of dignity, worshipped by many but forgotten today. Yet, most of us ignore the Earth we walk on, and are blind to the idea of deity and life force in Earth. We display something very close to a religious addiction to cars, to television, computers, gasoline, alcohol, tourism, beauty contests, sporting events, speedway races, and a host of other rather frivolous things. But few can raise their sights to realize the sacredness of the Earth beneath everyone's feet.

The idea that an ecocentric ethic and philosophy would tend to reduce the fragmentation in society has considerable merit. It bears more than a smattering of logical power and an equal amount of what could appropriately be called ecumenical power. There is much wisdom involved in appreciating what we have been given by the Earth, which is undeniably our life giver and life taker unless governments, industry, automobiles, aircraft, or other human innovations take it first. There is wisdom in having many humans employed with dignity in the care of this miracle ecosphere, which came into being long before bacteria invented fermentation, photosynthesis, and oxygen production by the plant kingdom. Only after the oxygen content of Earth's air reached 21 percent was it possible for a Johnny-come-lately species such as ours to exist.

Social justice issues, regardless of importance, cannot be resolved "unless the hemorrhaging of ecosystems is stopped by putting an end to homocentric philosophies and activities." Only then will there be reason to hope for the continuance of our species.

We pretend to high technology, but look around at the few remaining places that have not been brutalized by industrial society. The winds that greeted early visitors to this continent were fresh and nourishing. The waters were pure and filled with fish. The woods and marshes were alive with wildlife. Near where I live is a lake called Trout Lake, which experienced a mining boom in the early twentieth century. In those days a fisherman who fished on the lake to supply homes in that small community was known to throw back or even to kill forty and fifty-pound rainbow trout because people didn't enjoy eating fish that big.

What changed it all? What turned the fresh water into mud-filled streams in the spring, and caused the air over industrial cities to be choked with foul gases from the incomplete combustion of automobiles, and smoke from industrial enterprises that vent foul and often life-threatening gases into everyone's air? Is there a single name for all this venting of gases? I think of it as flatulence. It is the reckless use of machinery in a thousand forms, and is both visible and lung-irritating proof of the fact that our technology nowhere near matches Nature's technology. Nature produces beauty, health and integrated balance without working one extra shift, without going on strike for higher wages, and without fouling the nest of every living thing.

Social justice must involve the most respectable intellectual decisions ever made by our species: to curb our numbers and our expectations. We could offer full employment to the world by restoring hundreds of millions of acres of eroded land produced by improvident agriculture, reckless deforestation, and devastating wars. Our present, very lame media could be used to educate people as to their true relationship with the Earth. Our children could be educated to be good citizens of the Earth.

Soil at Risk

An interesting example of educational resistance to justice toward the Earth is the study *Soil at Risk.* This lengthy study was produced by the Standing Senate Committee on Agriculture, Fisheries and Forestry and the Private Legislation Branch. The study took years to complete and was initiated by Senator Herbert O. Sparrow in 1984.

The report started with the warning that, "Canada is facing the most serious agricultural crisis in its history and unless action is taken quickly, this country will lose a major portion of its agricultural capability."

Soil at Risk is filled with information that should be part of the educational heritage of every Canadian student. This assumption is based on the fact that although people are still interested in eating food, the study has been virtually ignored by our educational institutions.

The report states that between 1961 and 1976, Canada suffered the permanent loss of more than 5.5 million acres of rich agricultural land to urban use. This is a serious problem as "less than nine percent of Canada's land area is capable of being cultivated." We are not cherishing and protecting land needed for food production. Twice the report touched on education. To increase awareness in students, the committee felt "that the Provincial Governments (must) commit themselves to the introduction of soil degradation and conservation studies at the primary and secondary school levels through the addition of environmental courses."

(p.14) Later in the report, the suggestion was made that "environmental studies should be instituted as part of school curricula from primary grades upward." (p.32) There were others suggestions such as "proclaiming a National Soil Conservation Week; and running 30 second conservation advertisements instead of beer commercials during hockey games."

I had dinner one evening with Senator Herbert Sparrow, head of the committee that produced *Soil at Risk.* To a large extent he despaired of political and educational organizations unable to come to grips with the essential importance of soil and its need for care. The published fact that nearly twenty-five billion tons of soil erode each year throughout the world, does not impress people as much as the specifications of a new vehicle or the number of saves some hockey goalie makes in a single evening. The problem is not merely our diversion by trivia, but our preference for such diversion to learning about things that require either studiousness or unentertaining concern. Certainly there may be exceptions, but I doubt that more than a tiny percent of legislators could write a comprehensive and instructive paragraph about soil function.

The food people eat should come from healthy, well-cared-for soil. Industrial processes dependent on artificial fertilizers and pesticides seriously damage the integrity of soil. Nonetheless, when the products of digestion are borne to their destination by the blood, they are either broken down to supply energy (catabolism) or are built into new protoplasm (anabolism). Thus breakdown and synthesis are carried on at the same time and are kept in dynamic equilibrium. If nutrients are taken in excess they become stored as fats. The processes of catabolism and anabolism might symbolically be extended to soils and to the fertility of lands and the planet. When land is rent asunder to provide profit at whatever cost, it is treated catabolically. When it is nourished with composted manure and with other organic residues such as agricultural waste, leaves and grass clippings, these materials benefit soil by enriching it. When this material decays, the essential organic substance returned to Earth produces large populations of soil microorganisms, which are a characteristic of healthy soil. A cubic foot of good soil may contain a billion microorganisms. By analogy, this nourishment of soil is an anabolic process.

Because of John Q. Public's estrangement from Nature, precious soil is treated merely as an industrial substance to be bulldozed into piles or compacted by heavy machinery. Such human efforts are catabolic, and have been carried on for years. We childishly think of the use of large machinery as "macho," and it is part of our ignorance that we feel real concern for the land to be "womanish." While the terms anabolic and catabolic (building-up and tearing down) are usually not used loosely to describe a species such as ours, the choice of reductionism as a standard industrial concept shows not only very limited vision, but vividly demonstrates economic processes that are basically catabolic (destructive) toward the entire planet. The problem with devoted homocentrism is that it expands our state of denial by causing us to assume that the people who rule society know what they are doing, by pretending that the laws of Nature do not apply to humans as well as other animals, and of course by shallow governments refusing to come to grips with problems that must be addressed in an ecocentric manner that will protect Earth.

Consider one aspect of social justice — producing healthy food to feed the hungry people of the world. Before this issue can be addressed realistically we have to recognize two things. We are products of nature, designed to eat food produced by natural processes that sustain soil health. Industrial society attempts to shortcut Nature's ways and replace them with more financially remunerative, but unnatural ones. People must awaken to the truth that respectful treatment of Earth brings benefits in the form of properly grown food. These benefits literally "trickle down" in the form of healthily grown food for consumers. Consider this matter and you will realize that this is an example of social justice that has been sacrificed by eliminating small farms where true husbandry can be practiced. Thinking of E.F. Schumacher's book title, *Small is Beautiful,* it is obvious that food production should never have been allowed to become simply another profit-focused mechanical process.

This boils down to fundamental information that should be provided in schools. The reality should be taught that we live on Earth, are dependent on the health and wealth produced by the planet, and that we are proceeding recklessly to destroy our planet. Therefore we are moving along rapidly on the path taken by the dinosaurs. Students need desperately to develop more respect for Planet Earth than is being shown by today's society.

Teachers and parents have a serious job ahead of them. Awakening our political leaders to these facts is among the hardest tasks.

It is obvious we have a long way to go and serious changes to make. I am reminded of a story that appeared in the Lethbridge Herald on June 15, 1898. "A man who had some horses to sell wrote to a friend in Ottawa asking if they could be sold in that city. The friend replied, `The people in Ottawa ride bicycles, the wagons are pulled by mules, the street cars are run by electricity and the government is run by jackasses, so there is no demand for horses here.'"

~~~

[1] "The 14 Worst Corporate Evildoers," A Global Exchange Report, posted December 12, 2005, www.alternet.org/story/29337/

[2] Ibid.

[3] Ibid.

[4] "Corporate Social Responsibility Reports a Danger to Sustainable Future," *ecologist,* October 2006, p.009.

[5] Edward Goldsmith, *The Great U-Turn* (Hartland, Devon, 1988) p.183–217.

[6] Cecille Carroll (Paddy Carroll's daughter) "Cataline: Peerless Pioneer of the Pack Trains," *PLUS,* March 13, 1993

[7] Tom Morris, *The Stoic Art of Living* (Chicago & LaSalle, Illinois: Open Court, 2004) p.58.

[8] *The CCPA Monitor,* Ottawa, November 2006, p.30

[9] Sir Percy Sykes, *A History of Afghanistan,* vol.1 (London: Macmilland & Co., 1940) p.224.

[10] Robert F. Harrington, *To Heal the Earth* (Surrey, BC: Hancock House, 1990) p.21–22.

[11] Robert Rienow and Leona Train Rienow, *Moment in the Sun* (New York: Ballantine Books, Inc., 1967) p.78.

[12] *The Stoic Art of Living,* p.58.

[13] D. Olive, *Just Reward* (Toronto: Key Porter Books, 1987), p.23.

[14] Ron Graham, *God's Dominion – A Skeptic's Quest* (Toronto: McClelland and Stewart, Inc., 1990) p.368–369.

[15] Victor Hugo, *Les Misérables* (New York and Scarborough, Ont.: Nal Penguin, Inc., 1987) pp.1256–1275.

[16] *Philosophy, The Basic Issues,* Ed. By Klemke, Kline, and Hollinger (New York: St. Martin's Press, 1982) p.291.
~~~

7: The First Action Principle Defend and Preserve Earth's Creative Potential

This action principle has been overdue since the Industrial Revolution began. It does seem that industrial potential has affected human reason. The Luddites (1811–1816) and other groups that cautioned against mechanical innovations that would create unemployment were mocked and sometimes beaten and killed. Not enough people recognized the impacts industrialism would ultimately have. It was also impossible to foretell the grievous effects on health that widespread pollution would cause. Technology, given an inch, has taken many miles and has destroyed thousands of species of organisms, and is on the verge of destroying this planet. Governments, particularly in the US and Canada, seem numbed

and rendered ineffective by the need for immediate reduction of the abuses that have been committed against the ecosphere by rampant industrialism.

Today's plight is a consequence of the actions and mechanisms that foster the destruction of ecosystems. The Manifesto states, "As first priority, the ecocentric philosophy urges preservation and restoration of natural ecosystems and their component species." It recognizes the possibility of major disasters that might affect the planet. These possibilities include collisions with comets and asteroids, and must also include the likelihood of widespread tectonic activity that may result from changes that have been made in the Earth's strain field by mega-projects such as large dams and reservoirs constructed by humans.

The authors of the Manifesto advise readers that the evolving creativity of Planet Earth is a continuing factor if it is not interfered with by the extinction of species, the destruction of ecosystems, and by the poisoning of soils, air and water. If the industrialized nations are wise enough to act vigorously, quickly, and even dramatically to reduce both impact on Earth and the population level in a substantial manner, we might evolve toward greater sensitivity, compassion, and intelligence. For a solution, people must face squarely the underlying challenge to our intelligence and insight by transcending the multitudinous follies offered by life. Our response must include major reduction of the demands made on the planet. These would include limited use of all forms of machinery, especially those used for travel and others used for production of unnecessary paraphernalia. Two ends might be achieved. One would see a decrease in the emission of carbon dioxide, other toxic gases, and toxic metals. The other is that we might, as a consequence of inspired behavior, find underfoot the long abandoned trail to civilization. Greed and sophistication would naturally yield to higher aspirations. Attitudinally the suicidal, corporate "conquer the Earth" mentality would be dismissed as irrational. This attitude quite probably results from blindness related to the limited liability enjoyed by corporations. Yes, the nature of jobs would change, but the amount of work to be done is immense and realistic, rather than trivial.

An Example of Relentless Pollution

Change will entail substituting modest, but sensible, expectations for the grandiose ones that exist today. Most people know innately, but must fully recognize, that the cumulative effect of poisons in our environment has been a problem for decades. For example, an article that appeared in *The Sunday Times Magazine* (London) by Jonathan Raban in 1988 ("A Short, Sharp, Shallow, Spiteful, Sea") pointed out that the North Sea is dying by degrees. Recipient of man-made poisons from many industrial rivers of Europe — the Scheldt, Rhine, Meuse, Weser, Elbe, Humber, Tees, and Thames — the concentrations of chemicals in the North Sea were already so large that eaters of shellfish were consuming more than half the World Health Organization's maximum daily allowance of cadmium from this source alone. The Rhine River, then Europe's worst polluter was vomiting thirty tonnes of organohalogens and one hundred tonnes of toxic heavy metals into the sea every day. By the time it reached Rotterdam, it had picked up more than 50,000 chemical compounds in a million different combinations. At the time the article was written, the North Sea was receiving 300 million gallons daily of sewage discharged from Britain. Also, 30 percent of fish caught in the German bight were diseased. In the North Sea, 400,000 tonnes of oil spillage were occurring each year, and off the coast of The Netherlands toxic waste was implicated in high levels of disease in flatfish.

If we peruse data about the North Sea given both years earlier and later than 1988, we find that society is steadily marching toward disaster. The industrial world shows no willingness to alter its suicidal pace, and societies are supersaturated with politicians who do not act as though they understand that government is derived from the word "govern," meaning "to exercise continuous sovereignty over; esp: to control and direct the making and administration of policy. And to control, direct, or strongly influence the actions and conduct of." True governance involves sincere and superior application of public concern towards the wholeness of the planet and its cargo. In each and every country where serious pollution is occurring (which is almost everywhere), where rivers are carrying all sorts of dangerous material into lakes or oceans, where factory chimneys are belching out noxious fumes, where automobiles and other fossil-fueled machines are used in trivial ways, politicians should arise and begin governing in the sense of substantially reducing the pace at which we are destroying ourselves. There really isn't any other choice.

This principle of the Manifesto also states, "Actions that unmake the stability and health of the Ecosphere and its ecosystems need to be identified and publicly condemned." Let us apply this thought to the world's corporations, which continue to act as a serious handicap, even a menace, to caring for Earth.

Among the hazards being created by industry are many toxic materials that have invaded the entire world. In the US and Canada, the mining industry is accountable for nearly half of industrial pollution that is recorded. The nature of hard-rock mining often involves removing extensive amounts of surface soil, as in the case of open-pit mines where the ore deposit lies near the surface. When the ore deposit lies at depth, adits, tunnels and shafts must be excavated, and tailing piles result from waste rock removed during mining. Water pollution by metals occurs very frequently and toxic metals invade streams and the water table. When metals normally buried in the Earth enter manufacturing industries, they eventually become distributed throughout the world and invade the bodies of living organisms through soil, air and water. Zinc, mercury, cadmium, selenium, copper, lead, antimony, and molybdenum are among the toxic metals that have become widely distributed in soils, and many are also known to exist in the tissues of organisms.

Far greater than the conniving of those who seek absolute power is the organization of Earth itself. With bewildering numbers of inorganic mineral types and consequent specialized soil formulations, with multiplicities of climate conditions, and millions of organisms specially tuned to the ecological conditions in which they exist and thrive, the ecosphere expresses stunning sensitivities and resilience. Earth's ecosystems also have elastic limits from which they can recover from abuse without permanent injury. There are also plastic limits from which recovery is possible, although some distortion will remain.

How Not to Mine Gold

In regard to mining endeavors, the extent of destructiveness involved in removing minimal amounts of gold from ore appears irrational. In *Collapse,* Jared Diamond noted the process of cyanide heap leaching utilized in Montana to remove gold in low concentrations from the Zortman–Landusky Mine, which was owned by Pegasus Gold, a Canadian corporation. The amount of gold available amounted to one ounce of gold in fifty tons of ore. The ore was taken from an open pit and made into a huge pile placed inside a lined leach pad, and then sprayed with a solution of cyanide, an extremely poisonous compound. Cyanide bonds to gold as it seeps through the tailings. Once the cyanide has seeped through the ore and bonded with the metal, the cyanide-gold leach material passes into a pond, from which it goes to a processing plant that removes the gold. The leftover cyanide solution, containing other toxic metals in this instance, was sprayed on forest or rangelands for disposal or sprayed back onto the pile. Unplanned events took place. The leach pads developed leaks from the weight of ore and machinery that pushed the ore around. The pond overflowed during a rainstorm. Cyanide gas was formed inadvertently and almost killed some of the workers. Pegasus Gold went broke and the huge heaps, open pits, and ponds were abandoned. Leakage from the site will continue to occur until ended in Nature's own time. Forty million dollars, or more, will be paid by taxpayers for clean up of the site. It has become a visiting place for nations planning mining investments. It offers a display of "how not to mine gold."[1]

It is interesting and frightening to note the lack of public concern among businesses for the damage they do. In the instance of Pegasus Gold, although fined $36 million for its actions, it still evaded the costs of major and other unending consequences, which included the contamination of streams with toxic metals. In 1988, with surface reclamation done on less than 15 percent of the mine site, the company directors voted themselves more than $5 million in bonuses, and then transferred remaining assets to a new company they created, called Apollo Gold. State and federal governments adopted a new plan for surface reclamation, which would cost $52 million. Thirty million dollars would be taken from the fine levied on Pegasus, and US taxpayers would pay the remaining $22 million.

While this example is not characteristic of mining in general, it does indicate the widespread damage that can be caused by mining ventures.

Industrialized Farming Not Good Husbandry

Factory farming demonstrates a perverted concept of bigness that supersaturates the superficiality of the corporate worldview. *State of the World 2006* warns "Avian flu, mad cow disease, and other recent diseases that can spread from animals to humans are symptoms of a larger change taking place in agriculture." In spite of reports by the media, by veterinarians, by government officials, public health officials, and by farmers, these diseases are not natural disasters but are the consequences of shoddy practices undertaken by "concentrating meat production in the hands of a few large companies."[2] The serious psychosis of seeking excess wealth is evident, and this manic pursuit of riches has resulted in illness in animals caused by feeding them diseased or contaminated food. The result is contaminated animal protein in our food supply that poses threats to the entire world. Society has been the loser for hiding its head in the sand and allowing small farms and a saner view of life to be replaced by the single vision of industry.

Industrialized farming is the antithesis of good husbandry and of care for the soil base that underwrites animal and crop health. Crop rotation, the growth of diverse crops, restoration of soil through growing legumes, and the distribution of manure to cropland, have been abandoned to monocultures that rely on agricultural chemicals that destroy soils' natural health. It was recognized decades ago that monocultures are characterized by ecological instability since they produce ideal conditions for disease and crop pests. Diversity, its opposite, encourages stability, not only because certain crops are complementary to others, but also because diversity does not encourage high breeding concentrations of insect pests. In industrial farming, hedgerows and windbreaks are removed, eliminating natural conditions that support predatory insects, birds, toads and other consumers of crop devouring insects. Chemical poisons, which are broad-spectrum killers of various kinds, have produced unnatural conditions for the growth of food. Scientists have warned that faunal and floral simplification of soil organisms, many of which are microscopic components of Nature's decomposition system, is one of many threats to our existence. Of late years the poison problems have been deliberately increased by such techniques as killing potato plants with herbicides several days before harvest in order to facilitate mechanical reaping.

Unrelenting Deforestation A Health Threat

The authors of the Manifesto also recognize industrial forestry as inimical to world health. I recently saw a picture in a Vancouver newspaper of an enormous barge loaded with cypress logs intended for Japan. I was interested because Japan has had top-down management of its own forests for centuries. After suffering from the serious effects of early deforestation, a new outlook on forestry was proclaimed by the shogun in 1666 who warned of dangers of erosion, siltation, and flooding resulting from deforestation. He urged the Japanese public to plant seedlings and launched a nationwide effort to regulate forest use. This was a considerable and remarkable endeavor because although Japan has the greatest population density of any large, first world nation — 1,000 inhabitants per square mile of total area and 5,000 individuals per square mile of farmland — 80 percent of Japan's total area still consists of lightly populated mountain forests. These forests are extremely competently managed and protected.

Japan's insight regarding the importance of trees is admirable. While it is no wonder that it buys vast quantities of timber from Canada, which lacks the willingness to protect its own forest base, Japan apparently possesses far greater awareness of standing forests importance than most nations. Very little concern for leaving standing trees exists in Canada, where they are merely commodities for maximizing economic benefits. Wealthy nations remain fascinated with an economy that thrives on deforestation, and chauvinistically slaughters carbon-storing trees that are vitally needed to help stabilize world climate. Paradoxically, Japan prefers importing timber from other nations for its own use, but zealously protects its own trees. If Canada processed its own timber rather than shipping raw logs to other countries, it would be better for the world and for Canada. China also has realized that protecting its own forests is of utmost importance to their own survival. By enabling other countries to protect their own trees while we destroy our own, we trade the stability of Canada for that of other lands, and ignore the future of our own young people.

Japan, incidentally, imports quantities of wood chips from Australia, paying seven dollars per ton, and from them manufactures high quality paper and paper products that sell in Japan for $1,000 per ton. If this does not show that Canada, which has been shipping raw logs to Japan for quite a while, is missing the boat, it should at least give us pause for rethinking our priorities.

A scientific recommendation that all deforestation should cease was made thirty-five years ago, but the public does not find this sort of information in corporately controlled media. Canadians are convinced, by bombardment from mass media, that the quality of modern life is measured only by a rising Gross National Product. There should be a worldwide moratorium on forest destruction in order to put into effect the great changes in our societies that are necessary to avert the threatening disaster of total climate collapse, and the other unknown disasters that are likely to follow. The shocking truth is that in spite of scientific recommendations for a moratorium on deforestation, "Worldwide, governments pay companies $25 billion a year (in subsidies) to destroy the Earth's fisheries, and $14 billion to wreck our forests."[3]

Since North America was settled, unremitting deforestation has been occurring. For years it was assumed that forests constituted an infinite resource. Mechanization, corporatism, and unceasing greed have destroyed that illusion. Forests throughout the world are being obliterated. However there have been places such as Switzerland where avalanches and destruction of communities have led to protective laws. The Swiss Forest Act requires that 27 percent of each canton be left forested, and also requires permits for cutting trees, and calls for their replacement with other trees.

Of late there has been increasing awareness that forests are not merely a resource but are vast evapo-transpiration systems vital to climate stability. While dollars are insubstantial symbols of human profit and wealth, their acquisition has been sufficient reason to destroy forests worldwide. Now, though, it is being recognized that standing forests have immense value. The Chinese have noted that standing trees are of greater value than logs on the ground.

The Pembina Institute, an independent, not-for-profit environmental policy research and education organization, has asked the question, "Are other nations willing to pay Canada for preserving the boreal region's ecological goods and services?" The boreal region covers 58.5 percent of the country — 584 million hectares — extending from Labrador and Newfoundland to the Yukon. The boreal forest stores an estimated sixty-seven billion tonnes of carbon, equivalent to 303 times Canada's total 2002 carbon emissions, or 7.8 times the world's total carbon emissions in the year 2000.

Obviously the unleashing of carbon normally stored by forests has much to do with global warming. One estimate states that 65percent of carbon normally stored by forests is released to the atmosphere within five years after the trees are cut. Munich Re, one of the world's largest reinsurance companies, estimates that the global insurance industry may face costs of US$304 billion per year by the end of this decade. Based on their estimates, the value of carbon stored in the boreal "carbon bank" amounts to $3.7 trillion. An economy based on restoring Earth to health is our only hope for survival.[4]

Trees in Headwaters Reduce Flooding

We should not be devouring our forests with no respect for protection of our own water supply, forest soils, and flooding such as occurred when the Chilliwack River went over its banks after a ten-to-thirteen-inch rainfall in November of 2006. Headwaters of rivers should not be deforested. We do not need money that badly. And wherever headwaters have been deforested, they should immediately be replanted. One can see in the US that the lack of control of the Mississippi River is a result of deforestation of the headwaters of some 250 river tributaries, many huge rivers in themselves, such as the Missouri and Ohio. It is deplorable that in the Pacific Northwest of the United States hundreds of landslides occur annually. Ninety-four percent of these slides originate from logging roads and clear-cuts. Debris torrents from deforested watersheds caused billions of dollars worth of damage in a single year (1996).[5]

Regarding the Manifesto principle under consideration, the authors state that "the mining of toxic materials, the manufacture of biological poisons in all forms, industrial farming, industrial fishing and industrial forestry" should all be stopped immediately along with militarism. The Manifesto conveys that all of these activities are extremely dangerous to the health of the ecosphere. "Unless curbed, lethal technologies such as these, justified as necessary for protecting specific human populations, enriching special corporate interests, and satisfying human wants rather than needs, will lead to ever-greater ecological and social disasters."

Addressing the forestry issue in another manner, I would like to mention current logging activities within a few miles of where I live here in British Columbia. A neighbor of mine, now retired but a logger for years, recently visited an extensive, nearby clear-cut area and commented that for every tree that was taken "fifty were left in burning piles." Such a condition results from a number of factors. Small trees six or eight inches in diameter, and sometimes bigger ones, plus youthful seedlings are destroyed by the mechanical process of utilizing machines that should never have been built except for the fact they reduce the number of workers needed and make bigger profits. It is the land, planet, and ecosystems that suffer from such unconscionable methods of logging. In these days of increasing temperature and drier summers, the smaller trees should be left standing and would partially shade the land, thereby reducing evaporation. Small trees would carry on transpiration, which helps to form rain clouds. They also have some years of growth behind them and would speed up restoration of the forest. They would help to support wildlife and leave at least a semblance of the "multiple use" that forest industries often advertise. Also, by leaving small trees standing and thereby reducing waste, their roots would help to stabilize soil. The forest industry needs to develop a sense of social justice, and a much greater sense of actual respect for land. It should also realize that its negligence causes many people to seek medical attention for breathing impairments due to smoke from slash-burning fires. By eliminating this practice they would earn greater respect from the public, which marks slash burning as one of the great hazards of living in forested areas.

Eliminate Slash Burning

Elimination of slash burning by doing away with clear-cutting would also help Canada meet some of its debt to the ecosphere because millions of tons of carbon-dioxide would NOT be released each spring and fall as now occurs when slash piles are burned. By logging more sensitively and with greater restraint, enormous amounts of waste and other pollution would be eliminated. By now we should realize that every standing tree left on the land helps to perform at least some of the normal functions of forests. Clear-cuts, while endowed forestry instructors may favor them, represent an obscene method of shortcutting responsibility. *Atlantic Monthly* magazine, some years ago ran a lengthy article explaining how industry has invaded university education and tried to make universities servile institutions for their own purposes, so that we are now turning out professional foresters who are in effect trained by industry.

One of the apparent reasons for the current ruinous manner of logging is that the forest industry prefers to fall *all* trees in order to facilitate plantations. These plantations, however, do not have the benefits of biodiversity that natural forests possess. I am sure that the invitation to hubris undertaken by an unintelligent economy is starting to receive a grim response from the planet. Sadly though, modern industry continually displays the lack of wisdom that might be expected from refurbished chimpanzees that have gone far out on a weak limb.

Nature's Free Services — $33 Trillion Annually

Not to be ignored — though it usually is — the 1992 Earth Summit stressed the major importance of the natural environment in its enablement of a healthy economy. This idea received support in 1997 when economists quantified the importance of "Nature's services…things like the soil-holding capacity of tree roots and the flood protection offered by mangroves and other trees — at $33 trillion annually, nearly twice the gross world product that year."[6]

Other research sources have focused on the value of individual trees. "According to one calculation, a typical tree that lives 50 years provides, free, $196,250 worth of ecological benefits." When a tree is cut and sold as dead wood it brings revenue that is only 0.3 percent of the value it has contributed during its life as a standing tree. If left uncut, many species will continue to produce natural services of increasing value each year for centuries. "For example, a single 50-year-old tree has produced $31,250 worth of oxygen, $62,500 in air pollution control, $31,250 in soil fertility and erosion control, $37,500 in recycling water and controlling humidity, $31,250 in shelter for wildlife, and $2,500 worth of protein. These important values of trees were recognized long ago in the old English proverb: 'Those who plant trees love others besides themselves.'"[7]

Lackadaisical governments contribute heavily to all industrial displays of irresponsibility and short-term interests. I was with a forester one day who disapproves of logging methods currently employed. Looking at a site where soil damage was extensive, he ruefully commented that if there were more intelligent life in our capital city, things wouldn't be done as they are.

I visited the burning piles mentioned above and saw two immense piles burned just a week earlier. The number of trees burned would be in the thousands. Using the economic contribution figure of $196,250 worth of benefits contributed by a fifty-year-old tree, and dividing the dollar figure by fifty years, each tree contributes $3,925 in benefits annually. To be sure, as a tree grows larger it produces more annual benefits, but it still must reach the larger size by continuous growth from seedling to mature tree. Therefore, when a thousand trees are burned because of wasteful logging, the wastefulness amounts to a bit less than four million dollars, but that figure must be multiplied again by the average age of the trees involved. The benefits of trees, even in monetary calculations, will soon become astronomical. It should be realized that this damage to the world far outstrips the dollar figures that make up the gross national product. The public should have much more responsibility itself in trying to assure that as little damage as possible is done to land. This means that any trees not actually needed should be left standing out of respect for Nature. Sometimes I suspect that Nature is much more conscious of what is happening than most people believe. If human damage to land is ameliorated, the planet would benefit and people might reap unlooked for benefits (to their own psyches for example).

The only way to live successfully on Earth is to live with restraint and to conscientiously minimize impact on the planet. Many of us need to make the move from being avid consumers to being conservers. Attaining that realization even today would mean beginning to control population and reducing consumption of everything. A good start would be to focus particularly on reduction, or elimination if possible, of those actions and activities that introduce carbon dioxide into the atmosphere.

Would it not be to our advantage to relearn the enjoyment of being at home and of getting to know the local environment more thoroughly?

Educating Ourselves

The foregoing information offers educational material for all of us and could introduce meaningful questions regarding who we are, what our purpose in life might be, and how we might be able to live harmoniously and successfully. Schools especially have a sacred task to educate young people to study, and to seek the wisdom that is missing in our leaders and in most of the general population. People who have aspirations to become politicians also need greater knowledge. With the world showing signs of collapse, our present leaders are not even willing to throw out a life preserver by immediately taking steps to reduce carbon emissions now. It is very evident that there is need for some form of required study that will prepare individuals for entry into politics. There was a time when our population was smaller and our Earth more resilient. However, what we are now doing has caused the Earth to be in need of intensive care.

Of particular interest to Canadians thinking of change will be the knowledge that Canada has one of the highest numbers of vehicles per capita in the world and also one of the highest emission rates from automobiles. A 1995 study estimated Canada's emission of carbon dioxide from automobiles alone to be fifty-six million tons per year, which increases atmospheric carbon by fifteen million tons yearly.

As for storage of carbon in vegetation, it is more attractive to Canadian economists to put their reforestation activity into fast-growing plantations in southern climates. It is true that such trees grow faster and, thus, absorb more carbon in semi-tropical lands. It is also true that there has been much damage caused in Canada that is being ignored because companies can save money by storing carbon in the tropics and by ignoring the mutilation of over logged lands in Canada. To reforest enough to compensate for Canada's auto exhausts can be accomplished in southern locations by replanting 2.7 million hectares to sequester the fifty-six million tonnes of CO_2 produced annually until the trees are large enough for market uses. To sequester the same amount in Canada yearly would mean planting over eighteen million hectares at a cost of close to $13 billion, compared to $2.3 billion in the tropics. The preference of the business world is to save money, but the solutions of a "global economy" ignore ravaged land needing reforestation in Canada, and the consequent long-term effects on employment in this country. There was popular advice given some years ago re the fact that, while people should think globally, they should act locally. It does not make sense, and is of course highly unethical, for the Canadian government to circumvent necessary repair of obliterated forest in Canada by rationalizing a substitute, but supposedly adequate reforestation job on a different continent. Not many people would prefer to repair a Chinese rickshaw rather than their own car simply because the job on the rickshaw would be so much cheaper. Persons aware of the likelihood of a Gulf Stream collapse have probably read that this could lead to regional climate collapse in eastern Canada. For companies to gain carbon credits by reforesting foreign or tropical locations is merely letting our country go down the drain while industry buys the right to pollute even more excessively. The truth of the matter is that corporations should reinvest much of their profits in restoring land that has been maimed by them.

Peaceful Recovery of Character

If our leaders could be a bit visionary they might study the accomplishments of the Civilian Conservation Corps (CCC) in the USA in the 1930s and note that among other things done by the Corps, two billion trees were planted. It would probably be the best task ever assigned the Canadian Military forces if they were asked to "stand on guard" for Canada by planting billions of trees on lands degraded by massive deforestation. This would be implementing military forces already being paid by letting them plant trees in a battle against the solely extractive massacre of vital ecosystems. By some sort of spiritual osmosis, working actively at restoring damage we have done might cause our nation to wake up and abandon warfare as an economic plus. We must wage a new war against the callousness that has typified the mechanical juggernaut that is part of the quantitative measure of success we have adopted. Presently our GNP will leave us with much paper money, but no water, no trees, no wildlife, abominable temperatures, and brand new autoimmune diseases. There is much that could be accomplished if we use our intelligence to realize that money invested in restoring forests adds to the real wealth of the world.

Now is the hour. Public disapproval of the rape of the planet must be loud, clear, and continuous from citizens in all walks of life. Corporate subsidies should cease, along with continuous tax favoritism that has picked the pockets of the poor in order to satiate the rich. Militarism is a disease, and the awesome expenses of weaponry are absolute evidence that barbarism continues at a pace that suggests its only end will be near extinction of our species. Biological poisons of all sorts, and their acceptance, are illogical forms of mind warping. Industrial farming, industrial forestry and industrial fishing are shockingly wasteful and are carried on with disdain for popular concern, and with foolish fervor. We have become pointedly profane although we live in a sacred context.

Famed poet John Milton included this thought in *On the Lord General Fairfax:*

O yet a nobler task awaits thy hand
(For what can war but endless war still breed?)
Till truth and right from violence be freed,
And public faith cleared from the shameful brand
Of public fraud. In Vain doth Valour bleed,
While Avarice and Rapine share the land.

[1] Jared Diamond, *Collapse* (New York: Penguin Group, 2005) p.40–41, 456–57.
[2] The Worldwatch Institute, *State of the World 2006* (New York: W.W. Norton & Company) p.24–40.
[3] Norman Myers and Jennifer Kent, *Perverse Subsidies: How Tax Dollars can Undercut the Environment and the Economy* (Washington, DC: Island Press, 2001) p.14.
[4] The Pembina Institute, "Counting Canada's Natural Capital: Assessing the Real Value of Canada's Boreal Ecosystems" (Alberta: 2003) p.4.
[5] The Worldwatch Institute, *State of the World 1998* (New York: W.W. Norton & Company) p.26, p.194.
[6] *State of the World 2002,* p.5
[7] G. Tyler Miller Jr., *Living in the Environment,* Fourth Edition (Belmont, California: Wadsworth Publ. Co., 1985) p.178

8: The Second Action Principle Reduce Human Population Size

This is a recommendation that appears to fly in the face of human rights. Suggestions to curb our population or even reduce our numbers by a slight percent offend us immensely. We do know that we cannot add fish to a goldfish bowl every day or every week. We recognize that the goldfish bowl is limited in size or, in other words, is finite. We have a much harder job to extrapolate the goldfish bowl example and to realize that our planet also has limits on the number of people it can feed, house, and support in every sense.

For years the US government had a sign in one of its public buildings that showed the daily increase in the population of people on the planet. The sign was accompanied by a logo that read, "More people mean more markets." This of course had a business flavor, and there was not enough applied wisdom on hand to make governments and industry consider the finiteness of Earth. Businesses were interested in production and profit,

and the more buyers there were on the Earth the more goods could be marketed.

We are experiencing severe overpopulation, but government circumlocution offices prodded by industry pretend that population will stabilize at fewer than nine billion people by 2050. The ecosphere has had enough of us and there is a distressing starvation problem at present. Overpopulation and global warming are both critical problems now and offer strong evidence that we will have to act hastily or suffer the consequences. World population is increasing at the rate of seventy-five million people per year, which roughly comes to another million people every five days or 200,000 additional people daily.

In this chapter various aspects and implications of the world's contemporary population problem will be examined. Reluctance to do any more than postpone decisions and maintain the status quo is characteristic of today's world and the money-seeking forces of society. Figure things out for yourself. Think of the costs involved upon entrance of a new individual. Those costs vary from exorbitant expenditures in a wealthy home where the new arrival has to have the best of everything, to the new arrival in an impoverished home in a severely degraded land, an individual whose brief life may be spent in undeserved purgatory.

There is an interesting little story that describes the direction we must take to solve the population problems in the world. "New (mental) patients were put in a room with concrete walls and floors, and each was given a large mop. An attendant then would turn on a big faucet and go out, closing the door behind him. The insane would go to work with the mops. The sane would turn off the tap."[1]

Ted Mosquin and Stan Rowe, authors of "A Manifesto for Earth," wrote, "A reasonable objective is the reduction to population levels as they were before the widespread use of fossil fuels; that is to one billion or less. This will be accomplished by intelligent policies or inevitably by plague, famine, and warfare…Country by country, world population size must be reduced by reducing conceptions."

Choosing between the alternatives, reducing conceptions is the most humane choice.

A Look at Carrying Capacity

The mechanics of population increase, stability, and decrease are biological issues that may not be well known even though they should be a part of basic education. The number of organisms of any species that can be supported in a given area is referred to as its "carrying capacity." Carrying capacity is determined by the amount of food, water and other life essentials that are available. Even when a population has reached carrying capacity it will produce more young, but these will perish unless enough older animals die to make room for them. Numbers of most species are adjusted by starvation, disease, predation, climate extremes, and by competition for adequate nourishment. The intensity of these forces work together to establish the carrying capacity. Obviously changes in the number of animals that can be supported may occur as land becomes more or less fertile, or as climate extremes become more common.

The total number of a species that can be produced by a pair of animals is referred to as the "biotic potential" and the forces that restrict population increase constitute "environmental resistance." Given favorable weather and food conditions, some animals, such as grasshoppers or Mormon crickets, may hatch in enormous numbers, consume the available food supply and migrate to other locations. Plague populations of migratory Mormon crickets sometimes cross roads in vast numbers, leaving the roads slippery from the crickets crushed by passing automobiles. Some grasshopper species become migratory and fly to new locations. Their population may increase for a number of seasons until unfavorable weather conditions, disease, or, sometimes, human intervention reduce their numbers.

Nature has long been a reliable guide to the implicit "laws of the universe," and scientific study has made it possible for us to learn that we are not immune from these laws. Carrying capacity of various habitats is one of the easiest of such laws to understand. On Earth photosynthesis could be considered Nature's law for food production. It produces everything we eat (vegetation and meat) and enables Earth's own natural economy to be of stupendously greater importance than humankind's technological economy. Thanks to what is variously termed as eternal law, cosmic law or natural law, spinning Planet Earth with its iron core produces a magnetic field reaching from one pole to the other. This field shields life on Earth from injurious cosmic rays. Yes of course, there are many things we have learned about natural laws. For instance, wildlife biologists have noted that under-stocked habitats produce healthier animals, and that soils derived from limestone develop animals with stronger bones than do soils derived from more acidic rocks. It is not surprising that much natural law can be understood if one is a student of life. We all know about the law of gravity, and this seems to be about the only natural law most of us do know about. As long as humanity chooses to pay no attention to natural law, it will not be possible to adopt intelligent policies to curb or reduce population numbers of our species. This stumbling block to stabilizing human population exists as a result of the egocentric assumption that we are immune from the law of supply and demand.

History verifies that great numbers of humans have died from crop failures, diseases such as the bubonic plague, and from warfare. The decline of fossil fuel supplies such as natural gas and oil products will exert new pressures on food supply and transportation. There is overwhelming evidence that the amount of carbon we produce, the shortage of water we lament, and the declining productivity of fisheries and agricultural lands are significant signs of human overpopulation. Every time we create a new "must-have" our ecological footprint increases. We continue to invite catastrophic environmental resistance.

We cannot afford to delay control measures that might avert disaster. Many people await religious rapture and others rely on God to solve all problems. This seems puzzling in view of religious instruction that the evidence of divinity can be found by "observing His works" (which do permit population collapses). Such people should be made aware of the Biblical observation in Ecclesiastes (3:19): "For that which befalleth the sons of men befalleth the beasts; even one thing befalleth them; as the one dieth, so dieth the other; yea they have all one breath; so that a man hath no preeminence above a beast; for all is vanity."

Starvation and disease, which follow overpopulation among other species, are what we can expect ourselves from over-stressing the carrying capacity of the land. This is particularly pertinent today, for our huge population is a result of extraction from Earth of the hydrocarbon remains of generations of other organisms that preceded our own days on Earth. What was buried by natural processes has been made a threatening part of the atmosphere by our unreflective procedures. The productivity of the planet has been "forced" by energy-remains from past millennia. As profiteers from the coal-oil-natural gas reserves we have attained prodigious population numbers and lavish times for many, but this can no longer be sustained now that natural gas and oil production have peaked and will be in steady decline henceforth.

Since 1950 scientists have made studies on the carrying capacity of the Earth in relation to the number of people it can support abundantly. The optimum number for the human species was determined to be 500 million people, which is about one thirteenth of the number of people alive today. "A Manifesto for Earth" in establishing the number at one billion people or less, based that number on the size of the human population that existed before the widespread use of fossil fuels. Though many thousands of people starve to death daily, and such starvation has been going on for many years and is increasing, there has been little evidence that people in developed countries are willing to reduce their own lifestyles in order to prevent others from starving. And there is less likelihood that the problem will be ameliorated, since 200,000 more persons are being born every day.

At the present time, two countries, China and India, both exceed the total population of 1 billion that the Manifesto suggests. China's population is 1.4 billion and India has 1.1 billion people. The two nations contain 40 percent of world population; as many people between them as in the next twenty largest countries combined. As might be expected, both nations have severe ecological problems. China for example has only 8 percent of the world's water for 22 percent of world population. On China's seven main rivers, 412 sites were checked for water quality in 2004. Of these, 58 percent were found to be too dirty for human consumption. In India, only about 10 percent of sewage is treated, and urban and industrial pollutants are commonly dumped into waterways. China is the second largest emitter of climate-changing carbon at one billion tons annually. Its production of that gas, while very large, is still only one-seventh of what the US emits, and India is one-eighteenth as high.

Other serious problems exist. For example, in both India and China, cropland is losing productivity due to erosion, desertification, and other types of degradation. A 1997 study of degradation of land in Asia revealed that 44 percent of land in China and 50 percent of land in India was degraded to at least some extent. More serious degradation affected 17 percent of China and 28 percent of India. Land degradation is closely associated with population impact, and the 2005 Millennium Assessment expressed a "No" as to whether the world's ecosystems can stand increasing carbon emissions, additional forest loss, and extinction of species.[2]

Letting Nature be Our Teacher

It is possible that population problems are solved by some species that have retained instincts lost by humans over years of relative detachment from Nature's ways. Extensive studies by V.C. Wynne-Edwards of Aberdeen University indicate that various kinds of animals have developed social actions such as strong territoriality, which tends to stabilize population levels. Territoriality leads to reduction in the numbers of breeding pairs due to stress resulting from inability to find suitable nesting sites. We may think that songbirds sing in spring to entertain us, but song is serious advertisement of a suitable nesting site. The singers, usually male, invite sharp-eyed females to check out the attractions of the nesting site they protect with song and threats to other males. Birds typically fly from one boundary to another of the territories they have staked out for their prospective mates. Other animal populations such as foxes, deer, and rabbits, react to overcrowding by a decreasing rate of ovulation due to lowered secretion of sex hormones, or by resorption of embryos after they reach the uterus. It may be more difficult for human females to assess the territory offered them, since a male could possess a flashy automobile (which he might have bought on credit) and skills such as dancing, playing hockey, and being captain of the tennis team or the fire brigade, but have little to offer as an indication of his ability to provide food for a family.

There is an important "however" in the preceding paragraph. This is that when a population of animals is isolated and safe from major predators, as we now are, it loses the instinctive behavior that may protect it from the ultimate disaster that accompanies huge population increases. Humans are animals that have been protected from major threats of predation for many years. A large predator, such as a lion, may look majestic and even admirable in a zoo, but let word of an escaped predator from a zoo circulate, and ancient, suppressed fears will be felt by many people — dim vestiges of long-buried instincts. On the other hand, predators of a different sort lure unsuspecting people to financial and even lifetime disaster. Skilled business predators of many kinds manage to trap people with such bait as "no down-payment and easy terms," or when times get more desperate, with "zero-percent financing." Political predators lure voters by kissing babies, flattering the public, and by false promises called platforms, which have the purpose of securing votes.

In all cases the Latin motto *caveat emptor,* which advises, "Let the buyer beware," is a handy self-protective tool with which people should be armed as a partial guide to life. This is an instance in instruction when a teacher may pass on a long-standing caution that can help students to be prepared for ubiquitous, exaggerated advertising. Back a few years, before hands-on learning was in vogue, there was a popular term for learning by example. It was called "vicarious experience," and the general theme behind such instruction was that one did not have to place a hand on a stove to find out if it was hot. Today's disregard of vicarious learning can be lethal, as can be seen by car crashes caused by overconfident drivers and the popular attitude about driving that "there's nothing to it!"

A Look at the Kaibab Plateau

Let's learn something about population by what has happened to another species. The event is a classic of natural history. In 1906, with the best of intentions, US President Theodore Roosevelt signed papers to form the Grand Canyon National Game Preserve. The intent was to protect a healthy herd of mule deer from competition with herds of cattle, sheep, and horses that ranged on the ninety-square-mile Kaibab Plateau. The plateau had the apparent advantage of being isolated by the Grand Canyon on the south side, by other deep canyons to east and west, and by semi-arid land to the north. Before 1906 there was a population of about 4,000 deer in the Kaibab. The deer were primary consumers (eaters of vegetation) and they supported predators such as plains wolves, coyotes, bobcats, cougars and bears. All of these predators hunted other things also, but the deer were a part of their diet. In 1906 the area was declared a game refuge. Cattle were no longer allowed to graze there, and trappers and hunters moved in to get rid of the predators. Between 1906 and 1931, 816 cougars, 30 wolves, 863 bobcats, and 7,388 coyotes were destroyed.

The deer never had it so good. There were no enemies to fear and no cattle to compete for food. The numbers of deer began to increase rapidly. Within eighteen years, from 1906 to 1924, the deer population increased from 4,000 to 100,000. Hundreds of deer could be counted during a short hike.[3]

What then happened is no surprise. Any environment has its limits. The shrubs were over-browsed and started to take on a sickly appearance. Branches were taken off trees as high as a deer could reach. Seedling trees were also eaten. The deer started to look gaunt and starved — and then they began to die. Between 1924 and 1930, 80,000 deer starved to death. After 1930 they continued to die from the same cause and from diseases. The carrying capacity of the plateau had been exceeded; the environment had been damaged. When the food disappeared, death was the only alternative. Predators are part of a natural community. Destroying the predators made the Kaibab community unnatural, and it suffered terribly as a result. The Kaibab experiment is an example of the adage that "men may have good principles without having good practices." Observations were made of severe stress/shock behavior during the deer die-off period in 1926–27. A forest service worker stated that deer appeared to be dazed and in a state of shock because of malnutrition. They could be easily approached, roped, and led into camp.[4] Likewise the stress experienced in modern life today cannot be disassociated from population pressures people experience as they compete for the means to obtain all the good things which are advertised as not only desirable but necessary to our lives. This pressure is increased as people jostle for space on the highways, and in line-ups of all sorts from airports, restaurants, banks, and supermarkets to various sorts of public events.

Endangered Mountain Caribou

For all our human pretenses of management, and the lessons learned from the Kaibab Plateau, we display no ability to manage the planet's ecosystem. "Today, B.C.'s Conservation Data Centre shows 1,364 species are endangered or threatened in British Columbia. Not all of these are caused by logging, but whether it's over-fishing, fish farms, toxic waste, or other causes, the same formula has been applied: maximum profits have been pursued at the expense of undermining the conditions for life on the planet."[5] To help protect the disappearing mountain caribou in British Columbia, the government has asked the Mountain Caribou Science Team to come up with recommendations. One of the options strongly recommended by the team was that of killing predators which feed upon the caribou. This would include wolves, cougars, wolverines, black bears and perhaps even some grizzly bears. In addition the team speaks of cutting the numbers of those animals that may compete with the caribou for rangeland, such as moose, elk, and deer. There is also the likelihood that trapping will be utilized and, although it is not mentioned, current mentality would probably use helicopters for hunting down animals the "managers" choose to kill. Certainly the process of chasing down woodland caribou with helicopters in winter for tagging purposes has killed many caribou by exhausting them in the season when they need their food reserves most. Little mention in the plan is placed on conserving the last remnants of old growth forests in the province upon which the caribou depend almost exclusively for their winter range.

The best way to manage for caribou or other animals is to stop destroying natural habitats by so-called development, which wipes out these habitats through excessive forest eradication that involves miles upon miles of roads to facilitate forest removal. What needs to be managed is the planet-destroying logging and the spider web roads that leave animals easy prey for machine-dependent-hunters who sit in comfort on their ATVs or snowmobiles, with a rifle ready to kill anything that ventures into sight. There used to be millions of woodland caribou that ranged widely in many provinces of Canada, and many northern US States. (See Seton's *Lives of Game Animals.*[6]) We do not display any sense of justice toward a species that has been forced to the brink of extinction by forest destruction in British Columbia. We seem devoid of sufficient instinct to realize that destruction of mountain caribou moves us closer to self-destruction. We are connected.

How do we who stand on guard for Canada, compare with other countries? Here are some facts: China has 1,600 giant pandas — and 1.4 billion people; Africa has 3,610 black rhinoceroses; and East Asia has 4,500–7,350 snow leopards. Canada has only 1,670 mountain caribou. These are isolated in a number of small herds, and the British Columbia government has indicated that some of the herds will inevitably be rendered extinct by logging, which is allowed to destroy their natural habitat in old growth timber.

In general animals have reproductive potential sufficient to cope with the environmental resistance they meet in life. We are familiar with bird-nesting each spring. We recognize that the number of parents is usually two. We also know that some species of birds have more than merely the replacement number (two) of offspring. I recall that some years when we encountered a ruffed grouse on our back trail, she may have had ten or eleven young. It is a reminder of Nature's masterful design to see that the young, only a week old or thereabouts, have sufficient instinct to remain immobile when danger threatens. It is not unusual for the mother grouse to threaten us with attack or to attempt to lead us away by wing-dragging or some other diversion. The camouflaged young who remain perfectly still are virtually undetectable. Our own presence, to the grouse, is interpreted as threat by a potential predator. Adverse conditions such as predation, severe rainfall, an unusual period of cold weather, inadequate food supply, or disease, are factors which are all referred to as environmental resistance, and though grouse populations are cyclic, environmental resistance acts to keep grouse populations stable over long periods of time.

The average number of any species occupying a given area is referred to as the population density of a species. It is obvious that the human population greatly increased since humans gave up being hunters and gatherers and instead developed agriculture based on domestic animals and plants. Evidence indicates that no more than five million people were alive in 10,000 BC, and that human numbers increased to about 545 million by 1650 AD and gathered momentum steadily to the 6.5 billion people who now occupy the planet. Though population density is very high in large cities, the urbanites are as totally dependent on the productivity of the planet as are the people who live in rural settings.

Obviously there are lessons that must be heeded in the Kaibab plateau story, as there are lessons to be learned from the rabbit plague introduced to Australia, from the plague populations of rats that infest many cities, and from plagues of insects such as locusts, Mormon crickets, and other animals. Gardeners do not need to be reminded of the persistence of weeds such as field bindweed, chickweed, docks, ragweed, and other plants that are "growing out of place." It is interesting but sad to note how many individuals poison their lawns to keep out the bright flowers of dandelions and other plants that have become unfashionable. Actually, fresh young leaves of dandelions are tasty as part of salads or as cooked greens. Wonderful wine is made from dandelion blossoms, which are also edible and tasty, and on numerous occasions over the years I have pulled a few dandelions, roots and all, and let them dry atop a stump, for later use as a not-bad coffee substitute. Miner's coffee, this drink was once called. I have often thought that our fickle market manipulations might designate dandelions to be a prize-winning flower. If these plants were thus re-evaluated and sold for five dollars apiece, and also were advertised as the focus of a stylish, showcase lawn, docile consumers would probably jump aboard the new bandwagon. Thoreau spoke words to the effect that if the head monkey in Paris wore a purple shawl, all the monkeys in America would also feel the need to own one. Why not dandelions — emergency food for a while?

Wars Caused by Overpopulation

War is a form of extreme strife. It doesn't take much to move our species toward belligerent confrontation, and it requires even less to put people in a personal survival mode. Linda has relatives who live on one of the Gulf Islands off the British Columbia Coast. A few years back, the workers who run the ferry service to the islands went on strike and people feared a lengthy shut down of service. Her relatives felt that it might be a good idea to get a loaf of bread the day after the strike started. When they got to the one supermarket on the island, they found an unusual crowd of people with shopping carts, stuffing them with great quantities of items, and the shelves were almost bare. They were more than a little surprised at the apparent panic that had caused so many people to stock up in anticipation of a crisis. I have heard on occasion that a big sale featuring limited numbers of highly desirable items will produce an avalanche of customers waiting to force their ways though the doors when the store hosting the sale opens on day of the sale.

If relative turmoil can be produced by such minor events, just consider what excess populations will do to produce wars. Newspapers and magazines have already advised the public that wars will someday be fought over water.

In Rwanda a population increase from 1.9 million people to eight million people took place between 1950 and 1994. Firewood traditionally served as fuel for cooking, but as the population increased so did the demand for firewood. Forest growth could not keep up and by 1991 trees were only a memory, and as a result people had to use crop residues such as straw for cooking fuel. Normally crop residues had been returned to the soil for enrichment. Fertility declined from the loss of needed organic matter.

Other population problems contributed to the massive attack by Hutus on Tutsis, which caused 800,000 deaths, mostly of Tutsis. The scarcity of land to support the number of people also helped produce this crisis. One problem had to do with small plots of land owned by families. These were marginal for their needs. Average families had seven children and the family plot, already small, had to be fairly divided between the seven children. When the massacre took place, whole families were slaughtered in order that no survivors would be left to claim the family plots for themselves.

The type of problem Rwanda experienced is nearing a crisis in other nations where population growth is outstripping the carrying capacity of the land. A potential war over water exists in the three main countries in the valley of the Nile River: Egypt, Sudan, and Ethiopia. Population increase in these nations is great. Egypt's seventy-one million people in 2003 will increase to 127 million by 2050. Sudan's thirty-three million will be sixty million by 2050, and Ethiopia's sixty-nine million will become 171 million in the same time frame. Nearly rainless Egypt now gets most of the Nile water for agriculture, but Ethiopia has 85 percent of the Nile headwaters within its borders, and both Ethiopia and Sudan will need more water in the future. With the bulk of the water from the Nile now being used in Egypt, per-person income in Egypt is $1,300 per year, compared to only $90 per person in Ethiopia.

Throughout the world there is no solution to keep humanity going other than population reduction. If humanity procrastinates, the problem will probably be solved in the manner suggested by Malthus, namely by war, famine, or disease.[7]

Jared Diamond, in *Collapse,* gives historical background of cannibalism and other practices that have firm footholds in human behavior when starvation threatens survival. The population numbers given for India and China give evidence of already overpopulated nations that are still growing with catastrophe in sight. Since solutions other than the Malthusian ones rest on political decisions and actions, it is folly that the search for international solutions to population growth is not given emergency status without delay.

How Iran Reduced Its Population

Could government efforts at population reduction work? They have worked in Iran and could work here. Family planning programs had been put in place in Iran by the Shah in 1967, but were put aside by Ayatollah Khomeini in 1979. During the period of 1980–88, a war with Iraq was occurring. Asking the public for large families, to provide his target of twenty million soldiers, Khomeini succeeded in pushing Iran's population growth rate to 4.4 percent per year. The population increase produced overcrowding, land degradation, and unemployment.

In 1989 Iran's government reversed directions and returned to a family planning program intensely supported by several government ministries: health, education, and culture. Iran Broadcasting was used to increase public awareness, and television was utilized to provide information on family planning. Religious leaders were urged to crusade for smaller families. Fifteen thousand health houses provided family planning methods and health services for rural people. Iran became the only nation that required couples to take a class on modern contraceptive measures before marriage. Male sterilization was recommended, and all forms of contraceptives and sterilization were free of charge. The main thing is that the effort worked because family size dropped from seven children to below three, and in seventeen years Iran cut its growth rate in half. Its growth rate by 2001 was down to 1.2 percent.[8]

Education is also a solution to overpopulation, as the education of girls has shown. In every study undertaken, it has been proved that the higher the education of the females of any population, the lower their fertility rates fall.

The doubling time for a population can be calculated by dividing the growth rate into seventy-two. For example, if the growth rate were 3 percent, the doubling time would be twenty-four years. If it were 2 percent, the doubling time would be thirty-six years. But the fact is that we need to move in the other direction, not toward doubling populations, but toward decreasing them.

This is particularly important now, since major nations such as China and India continue growing, and these countries are producing environmental refugees that will pour into countries both legally and by stealth. Canada, while indeed roomy, has vast amounts of land that cannot support populations. It may soon be necessary to restrict environmental refugees from entering host nations unless they are sterilized as a condition of entrance. Note that many people seeking refuge are young and in their childbearing years.

Immigration levels can easily become strained. In the US the National Origins Act of 1924 cut immigration quotas to 2 percent of each nationality that was recorded in the 1890 census.[9] However, Canada might be prompted, particularly by corporations, to allow in more and more people simply because they become customers for over-valued business, and cheap sources of labor.

The pressure of the numbers of humans has already begun to deeply undercut the ecological stability of Earth. The economy has clearly violated safety limits that should be observed for protection of climate, humidity levels, soil and stream stability, temperature and wind control, and many other needs, including carbon storage. Animal studies indicate that the healthiest organisms are those that live in habitats that are not crowded.

What about Canada's Population Problems?

One of the best studies I have seen on a population policy for Canada was written by J. Anthony Cassils and was published in *Proceedings,* a quarterly report of the Canadian Association for the Club of Rome (CACOR). Cassils stresses the need for a global approach to overpopulation. He offers a sensible series of steps that should be followed to develop a Canadian population policy.

The first step would be for the government of Canada and its provinces and territories to study and develop a long-term carrying capacity for humans that would also take into account the needs of other species and the biodiversity required to perpetuate a stable ecosphere. He suggests a need for every country in the world to calculate a carrying capacity of similar nature, but does not believe that we should delay our own efforts waiting for an international accord to be reached. The carrying capacity determined would include the needs for sustenance that take place within and outside the nation, thereby realistically considering exports, imports, and the assimilation of wastes. An understanding must come about that we are but parts of an integrated ecosphere.

From such study the national, provincial and territorial governments should provide leadership for a comprehensive population policy for Canada. Cassils notes that most politicians avoid confronting the population growth issue because of criticism they might receive and because of possible damage to their careers. His contention is that politicians must move from a merely reactive form of governance to a new more assertive form. In short, real leadership is needed.

The long-term carrying capacity of Canada, once ascertained, should provide an objective base for an ecologically sound and socially just population policy. He states that: "The Government of Canada and the Governments of the Provinces and Territories should actively promote the development of the branch of ecology that looks at the carrying capacity of the natural environment and at the ecological footprints of various communities, cities and countries. Too much damage has already been done and restoration to former levels of productivity should now be a priority."

The various governments should educate Canadians about the "acute dangers of overpopulation," and should encourage public support and participation for a population policy. Public insight should also be sought and employed in creation of such a policy.

The various levels of government should take steps to slow population growth until carrying capacity figures are available and we can utilize them. Meanwhile, increased international aid should be given other nations willing to develop policies to lower fertility but needing assistance. Without immigration, Canada will have slow natural growth for the next fifteen years (Statistics Canada), and reducing immigration to levels prior to those of 1989 could slow population growth. Cassils notes that the Mulroney government, in an effort to win immigrant support, raised immigration levels in such manner as to increase Canada's population by one percent a year. According to Cassils, "Since 1990, recent immigrants have been doing substantially less well than the average Canadian, when a generation or two before, they did considerably better." This evidence suggests that Canada's carrying capacity may have already been reached. He states, "the current level of migration to Canada, as a percentage of population, exceeds that of all other developed countries." He also comments that parents who wish to have smaller families are stabilizing Canada's population.

Cassils feels that individual nations must reduce their own growth rate and not rely on other nations to welcome their excess population as immigrants. He points out that a nation that reduces its own growth rate should benefit from its own foresight and "must not be invaded by illegal migration or by military action. Nor should they give in to the advocates of growth within their respective countries by allowing massive immigration...Countries that allow their populations to rise beyond carrying capacity must face the results of their actions or inaction as this will make evident to them very rapidly the need to change their habits and cultures."

Mr. Cassils also reminds readers, "The good news is that populations that grow exponentially can shrink exponentially. A few generations of below replacement fertility could reduce the global population to sustainable levels." He also comments that those who advocate "perpetual growth bemoan lower fertility rates as they rush to lay waste what remains of the living Earth."[10]

Suffice it to say that the author of "Strategy for a National and International Population Policy" is a highly qualified individual whose thoughts were prescient but have now become extremely pertinent and necessary in a world wherein equivocation and procrastination — noted political characteristics — can no longer be afforded.

[1] William Vogt, *People!* (New York, 1961) p.169.

[2] *State of The World 2006* (New York: W.W. Norton & Company) p.1–15.

[3] Durward L. Allen, *Our Wildlife Legacy* (New York City: Funk & Wagnalls Co., 1962) pp.234–235.

[4] Ibid. p. 256.

[5] "How the Government of British Columbia is Killing Endangered Mountain Caribou," *Valhalla Wilderness Watch,* September 2006, p.6.

[6] E.T. Seton, *Lives of Game Animals,* Vol. III–Part I (Boston: C.T. Branford, Company, 1953) p.60–65.

[7] Lester R. Brown, *Plan B* (New York & London: W.W. Norton & Company, 2003) p.101–108.

[8] *Plan B,* p.178–181.

[9] James H. Kunstler, *The Long Emergency* (New York: Atlantic Monthly Press, 2005) p.205.

[10] J. Anthony Cassils, "A Strategy for a National and International Population Policy," Can. Assn. for the Club of Rome, *Proceedings*, January 2006, p. 17–26.

9. The Third Action Principle: Reduce Human Consumption of Earth Parts

A certain amount of suffering and enduring adversity will be entailed in returning society to a sensible economy that functions with deep regard for the health of the planet. Governments and corporations will have to reverse roles, with governments regulating corporations and corporations acquiescing to measures that will put the integrity of Planet Earth as the first priority. Then and only then will it be possible to start in earnest to "Reduce Human Consumption of Earth Parts." The need for that reduction is expressed in the words "The chief threat to the Ecosphere's diversity, beauty, and stability is man's ever-increasing appropriation of the planet's goods for exclusive human uses." The Manifesto authors point

out that "this steals the livelihood of other organisms." Furthermore, they justly contend that it is "morally reprehensible" for us to chauvinistically claim that it is all ours.

It has been widely recognized that greed has been glorified in modern society. It has been promoted from being a serious sin to the position of an admirable virtue. The carefully restricted diamond market, as an example, enables the wealthy of the world to store these agreed-upon treasures quite compactly and convince themselves that diamonds are of enormous value. Nobody seriously asks, "What for?" To be sure, minerals of various sorts form beautiful crystals of many shapes and sizes. Personally I think that the rhombic dodecahedron of garnet, if blessed with good color and sparkle, outweighs a diamond for beauty (particularly almandine garnet of jewelry grade from Madagascar and India). But all of these treasures are creations of the unsurpassable artist, Mother Nature. Diamonds are a crystalline form of the element carbon, and there are multibillions of them in volcanic necks. Diamond merchants create their value, making sure that the supply of desirable stones of approved size and sparkle is never allowed to exceed the demand. Fresh air and clean water, however, have real value that mankind abuses with insane fervor. It would be worth mountains of diamonds if air and water could be returned to their natural state, but we are swapping our inheritance for a mess of pottage.

The manipulation of the diamond market, keeping diamonds at high prices to convince people of their rarity and preciousness, is one of the accepted business practices perpetuated in society. Enough diamonds could be produced to make them available cheaply. Currently the illegal diamond trade in countries such as Angola, Sierra Leone (and others) pays workers less than a dollar a day, plus a bowl of rice, to work waist deep in water to mine diamonds used to buy armaments and support civil conflicts. These diamonds are often referred to as blood diamonds. On occasion, individuals suspected of having swallowed a diamond have been eviscerated in order to recover one of the gems.

We do little to protect our integrity. Much money may be spent, hours of saturated and quasi-pleasure may occur, considerable sickness may result, and a total failure of health may come from consuming alcohol in its various tempting garments, but invariably the use of alcohol destroys cells. We likewise take much pleasure in eating various foods, forgetting that almost all of them contain pesticides capable of producing many diseases. If pesticides were abandoned and more people were put to work on farms to cultivate, weed, and work at less poisonous methods of insect control, no pesticides would be needed. The workers employed in producing pesticides could be alternatively hired to work in agriculture. Today we tend to spurn the thought of working on the land although the greatest of the miracles that surround us have much to do with soil. Nathaniel Southgate Shaler, a Harvard University professor, described the role of soil in 1905 when he made the analogy that soil constitutes a living placenta surrounding our planet and is eternally pregnant with new life.[1] In *Organisms and Ecology of the Soil,* authors Nyle C. Brady and Ray R. Weil observe that a single gram of soil may contain up to 10,000 species of microorganisms, an amazing indication of species diversity.[2] These figures suggest that pesticides and other agricultural chemicals could cause severe faunal and floral species reduction, which would seriously impair or destroy essential life functions. As a species we know so much and yet know so little and our willingness to take risks is very apt to become calamitous. There is inherent nobility in working with and caring for soil, and life close to the land has long been considered admirable.

Someone might ask, "What? Do you want to go back to the Dark Ages?" A suitable answer might be that no matter how many kilowatts of electrical energy are used, we may be living in the darkest of all ages. Perhaps we are at a time when to step backward is to go forward. Life would become simpler, for example, if people lived close enough to their work to be able to walk or use a bicycle to get to the job. Recognizing this fact, the Swedish retail chain Ikea showed its concern for global warming at its recent Christmas breakfast by giving all 9,000 of its employees a free fold-up bicycle, and offered a 15 percent subsidy on public transport.[3] This was a generous and sensible gift, one that other corporations could emulate.

As humans we are free to make the choice to turn outward and live in harmony with other creatures, intellectually and spiritually as well as physically. Instead the majority have turned inward to materialistic concerns, and chosen to develop an uncompromising form of species arrogance. For all the macho ridiculing of birdwatchers and Nature lovers, such outwardly concerned individuals have attained a level of maturity that has been rejected by those who feel it is a human right to trample all over other living beings. It might help us to consider that one of the meanings of wild is "free"; and we fail to have real democracies because we sell for a pittance our right to think for ourselves. We seek entertainment via television for example, but those who advertise and study human nature have learned to profit from the fact that a bit of low-caliber entertainment can be mixed with indoctrination. What it amounts to is that when you give up thinking for yourself, there is always someone happy to step in and do your thinking for you. Encouraged to let others run the show, most people have not only given up their freedom but have gracelessly turned away from the reality of our very beautiful planet. As a consequence we have surrendered to the serious sin of alienation from Nature.

Consuming the Planet

Suppose that we reject the idea of being alienated and we want to reunite with the real world of which we are one part. We would need to review our actions and options, our possessions and our wants. We might decide, for example, not to buy a new car, or to drive as many miles as usual, or to be a tourist to support airlines and tourism. Much of the solution to problems rests on governments whose penchant for international affairs causes them to ignore the public, the planet, and Nature's own economy while strongly supporting the philosophy that the "business of Canada is business." This is totally wrong because the first item of business should be the health and integrity of the planet that underwrites all life. Government, however, can only be awakened if many people make it uncomfortable enough to stir from its sleep.

Two things must be done at once. One is to stop over-consuming. Second, tell government that we are reducing consumption because we are endangering the planet excessively (which means damaging ourselves also), and we want the health of the Earth and its populace to become the first priority of government. With the ship of state leaking badly, it is folly to sell life jackets in preference to repairing the ship's hull. Give up things such as that new shed you have thought about as a storage place for the lawnmower. Get an item repaired rather than buy a new one. Press governments into making it mandatory for suppliers of the throwaway electronic industries to take back and re-use or recycle outdated equipment. Take a walking holiday rather than an automotive tour. There is an old Japanese saying that "the day you cease to travel you will have arrived."[4] Share vehicles for driving to work. Definitely compost food wastes, leaves, and grass clippings rather than plastic bagging them and hauling them to the dump. Plastics have become such a problem that recently a filter-feeding, minke whale was washed onto the shore in Normandy and 800 pounds of plastic bags and other packaging were found in its stomach![5]

Turn off lights you are not using. Plant a garden. Eat a bit lower on the food chain, more vegetables, less meat. Learn more about Nature and our relationship to it. Escape into Nature on occasion and find a log or a rock where you can think your own thoughts. Reuse, recycle, and repair. Set a definite goal for yourself such as reducing your own driving by fifty percent. Mahatma Gandhi, who died in 1948, once said that there is enough for everyone's need, but not for everyone's greed. However, with six and a half billion people now on the planet, this statement is questionable today. We are at the point where we must examine our personal definition of "need" each time we make a purchase. If what we need today is going to end up as garbage tomorrow, then those so-called needs must be relinquished for the sake of a decent future for our children and grandchildren. The inability to control appetites, for example, has reached such an alarming point that according to Statistics Canada's Community Health Survey in 2004, nearly 23.1 percent of all Canadians aged eighteen years or over — that's five and a half million people — were obese, while an additional 8.6 million were overweight. In Europe 31 percent of all adults are overweight. In the United States the number of overweight adults is 61 percent.[6] During the 1990s, 10 percent to 40 percent of people in most European countries entered obesity and in the United States 50 percent reached that stage. The causes are policies that make sugary and fatty foods cheap and plentiful while we perform less and less physical activity. As well, the risk of overweight people developing diabetes or heart problems is very serious. Walking a couple miles (a few kilometers) each day would be an important remedy. One thing that should be reiterated in the context of consumption is that the very best way to reduce it is to decrease the number of consumers. Population growth is an enormous threat to this world in which governments and industry almost beg people to increase consumption, travel and trivial pursuits.

The point has been reached at which all aware individuals know that steps must be taken to reduce consumption without delay. This involves everything from eating lower on the food chain, to reducing carbon emissions. Car sharing is a good example of how carbon emissions are being reduced. This innovation began in Switzerland in 1987 and was underway in Germany in 1988. By June 1996 there were seventeen car-sharing programs in the US with 101,993 members sharing 2,558 vehicles (U. Cal., Berkeley data); and there were eleven Canadian programs involving 15,663 members and 779 vehicles. In Canada a small or large car can be used for a four-hour trip for prices ranging from $18 to $34. The individual who rents the vehicle must pay for gasoline. The fee covers maintenance and insurance. In general the services are arranged by cooperatives. Obviously this idea falls in line with using fewer Earth parts, and certainly not everyone in a city needs to buy, house and pay other costs of a private vehicle. Businesses can be part of the solution too as indicated by a recent action of Citzens' Bank in Calgary that donated a 2005 Toyota Prius to Calgary's carsharing co-op.[7]

The First Deadly Sin

A narrow interpretation of the well known seven deadly sins named by Pope Gregory the Great in the fifth century, would enable us to understand that we are classic violators of the first of the of the deadly sins, namely pride. Daniel Defoe described pride as "first peer and president of hell." This first sin can be seen in the way foolish people who have amassed wealth and power refer to themselves as "elite." Such a self-deception represents the epitome of pride. Also our self-concern takes up so much of our thought and interest that we can hardly understand that there are other beings, whether they are large animals like deer and elk or small ones like chipmunks, which also share the Earth with us. We must make provision for others by leaving natural areas free from the depredation of progress.

We keep moving in the same direction, blindly committed to homocentrism. We must begin to exhibit care for other forms of life. Our species self-isolation has enabled us to impoverish the context of the ecosystems around the globe. We realize the pain involved in giving up our integrity and should therefore be aware that disrupting the integrity of ecosystems weakens their stability. Extreme homocentricity increases our need for constant stimulation and contributes to dependence that reduces self-reliance. Some of the popularity today of visiting gurus and retreat centers may be attributed to the underlying need that many who are harried by today's hectic pace feel a need for periods of meditation, prescribed silence and immobility.

We are Spiritual Beings

I have been impressed with the "mystic vision" expressed by Nobel Physics winner Erwin Schroedinger, and recognize his inference that a mystic vision can be personally experienced in a natural setting. He concluded his nearly poetic reflection with this thought: "Thus you can throw yourself flat on the ground, stretched out upon Mother Earth, with the certain conviction that you are one with her and she with you. You are as firmly established, as invulnerable as she — indeed, a thousand times firmer and more invulnerable. As surely as she will engulf you tomorrow, so surely will she bring you forth anew to new striving and suffering."[8]

Science, with its dependence on measurement and objectivity, does not venture beyond the material aspects of life, which it keeps as its domain. It does not accept individual speculations in religious or other metaphysical concepts. Yet, many scientists, including Alfred Einstein, Max Planck, Werner Heisenberg, Prince Louis DeBroglie, and numerous others have postulated that there is more to life than we are capable of understanding.

DeBroglie, a Nobel Laureate in quantum physics, in his book *Physics and Metaphysics* found himself in agreement with French philosopher Henri Bergson, who expressed his own conviction that humanity has more fossil fuel, hydroelectric, and atomic power available, than it could properly control. Like Bergson, he realized that "our mechanisms demand a mysticism." DeBroglie reiterated Bergson's call for a "supplement of soul" that would enable us to control the tiger that we have by the tail.[9] What DeBroglie's observation amounts to is that to solve the problems of our times we must reach deeply inside ourselves to find the essence of our being. To me, this means realizing that at the core of my being I am a fragment, a bit of Nature myself. Over the years I have sat on a lot of logs and boulders and pondered the idea of the meaning of life. I see this search also reflected in the words of the most eloquent of famous physicists, Sir Arthur Eddington, who wrote an essay, entitled "In Defence of Mysticism," in which he stated, "Whether in the intellectual pursuits of science or in the mystical pursuits of the spirit, the light beckons ahead and the purpose surging in our nature responds."[10]

Spirituality is a term that is difficult to define. As Fritjof Capra explains in *The Web of Life,* "Ultimately, deep ecological awareness is spiritual or religious awareness. When the concept of the human spirit is understood as the mode of consciousness in which the individual feels some sense of belonging to the cosmos as a whole, it becomes clear that ecological awareness is spiritual in its deepest sense."[11] What Capra says basically is that deep awareness of ecology is consistent with spiritual interpretations of many religions and of the beliefs of the North American Native people. What it boils down to is that in our deepest awareness we belong to something that is variously defined by different traditions.

In my own life I know that I have spent more than forty times forty days in remote places. When I think about spirituality I am reminded of when I was a young person and several of us would fish for eels at night. We all learned from experience that it is just about impossible to hold an eel, that's how slippery they are. Spirituality, I think, is something like that. When you try to grasp it, you find it intangible. But in wild places, the high mountains, along streams, in forests, by alpine lakes, I am pervaded by spirituality. There is an omnipresence about it that can be sensed but not described. All over the world it seems that religions of various sorts try to put a leash on that which cannot be leashed. This is just a personal thought from someone who understands his own unimportance.

What does seem significant to me is that so many highly intelligent people who work at the leading edge of culture recognize and publicly state that we have to reach deep inside ourselves to grasp the deepest truths.

They say, and we know inside ourselves, that the mind should not be fed by shriveling the soul. Our materialistic society tempts us to reduce or eliminate decency from many of our deeds. At present we are so devoted to the experimental quest that we refuse to cease the production of dangerous goods that threaten the stability of our ecosphere and even our existence. The mania for wealth and power over Nature as well as over other nations drives us in vicious ways. For example, it is well known that land mines from past world wars continue to kill and maim people in various nations. It has been estimated that random distribution of mines throughout the world has killed or maimed one million people since 1975 — and 80 percent of them are civilians.[12] In spite of that knowledge we now manufacture even more sophisticated mines and have developed a new horror called cluster bombs. It is not unreasonable to wonder at human sanity, and particularly at the vast extent of human sadism. When we see that today more casual acceptance and implementation of torture takes place, one realizes that the human is only marginally a humane creature. One may also look at the ritual use of biocides of all sorts to realize that whatever ill fate approaches our species may be well deserved. This is perhaps the reason why a "supplement of soul" has become an absolute necessity for a very troubled world and for our own species, which is rapidly progressing toward the lowest common denominator.

The famous historian Arnold Toynbee, states that ever since man became conscious it has been his objective to attain mastery over the physical world, yet that absolute mastery over the biosphere could well lead to man's own undoing. He cautions, "Man's other home, the spiritual world, is also an integral part of total reality; it differs from the biosphere in being both non-material and infinite; and, in his life in the spiritual world Man finds that his mission is to seek, not for a material mastery over his non-human environment, but for a spiritual mastery over himself."[13]

Just as Henri Bergson suggested, we humans have obtained more forms of energy and technical skill than we can handle wisely. Again realize that we are foolishly egotistic in calling our manufactured items "high technology." Take a close look in a mirror or look at a cat, dog, mouse, bird, or insect and you will be witnessing higher technology by far, than anything we can contrive. We have been grievously immature in setting out on a quest to subdue all Nature, to "conquer Nature" no less. Modern mechanized deforestation is an example of throwing caution to the winds. As long ago as 300 BC, Plato, in his *Laws,* urged that there should be control of deforestation. Before the time of Christ, the Chinese philosopher Lao Tzu identified the sacredness of Earth and the restless nature of man, saying:

> For the world is a sacred vessel
> Not made to be altered by man
> The tinker will spoil it,
> Usurpers will lose it.[14]

In our own recent history we have seen much evidence that human haste to develop new products has been found unwise and dangerous for our species. Albert Einstein, having seen his mathematical genius warped to make nuclear weapons, instructed his biographer, "Tell the world that if I had foreseen Hiroshima and Nagasaki, I would have torn up my formula in 1905." Another man whose foresight failed to inform him of what he learned by hindsight was Admiral Hyman Rickover, who directed the US nuclear sub program for forty years. Late in his own life he told a Senate committee that if he had his hand on the tiller he would sink the submarines he was responsible for and try to eliminate dependence on nuclear power.[15]

There is strong evidence that we, as a species have sinned in haste; but further evidence in the erratic nature of our climate and the continued orgy of fossil fuel use indicates that we may not have enough time left to repent at leisure.

In our limping education we must now dust off and implement wise words from "Science for Every Student," a guideline for educators from the Science Council of Canada, and also a timeless comment on education: "Since the time of ancient Greece, the two basic objectives of a liberal education have been the intellectual and moral development of the student. The Science Council reiterates the importance of these objectives for contemporary science."

Regarding the ancient wisdom that is part of human heritage, the statements following this paragraph are symbolic of values passed on to us that have been effectively obscured. Carefully controlled, modern society has conveniently renounced such thoughts as being out of step with present day technology and as a superfluous stumbling block in the pursuit of progress. A return to the accumulated wisdom of the most respected of our predecessors is nothing less then a light at the end of the seemingly opaque tunnel along which society is hobbling. The people who uttered these statements would, and presciently did, consider our present lives of haste and waste as a form of madness.

The Socratic Injunction — KNOW THYSELF

The Aristotelian Injunction — REALIZE YOURSELF

The Christian Injunction — REPENT AND RENEW YOURSELF

Buddhist Injunction — RENOUNCE YOURSELF

In studying the past, we will also learn that the Golden Rule has been freshly discovered as it was stated by one after another of the ancient religions and philosophies. The late Harvard University paleontologist Stephen Jay Gould (1990) in his column in *Natural History Magazine* argued that it would be enlightened self-interest on our part to adopt the Golden Rule as the major principle in our relationship toward the planet.

Pointing out that the planet holds all the cards, such a principle, he said, "would be a blessing for us, and an indulgence for her." As he stated, "We had better sign the papers while she (Earth) is still willing to make a deal." But, he continued, "If we scratch her she will bleed, kick us out, bandage up, and go about her business at her planetary scale."[16] Stan Rowe's view on the Golden Rule was that "The Golden Rule should logically be wholly subsumed by the emerging, more inclusive 'Green Rule.'" Realizing that we now know more about the ethical importance of the Earth and its ecosystems, a Green Rule would not only include our relationship to one another but to Earth which is home place for all of us.

For us to attempt to restore the spiritual quality of life, and reject the moral reprehensibility "of the homocentric view that humans have the right to all ecosystem components — air, land, water, and organisms," we will have to move from the quantitative goals now used to measure life. These goals need to be replaced by qualitative ones.

The Awakening

To accomplish the change needed on a large scale will necessitate putting much pressure on governments that are only as sensitive to public opinion as is absolutely unavoidable. It is a truism that the masses of people can rule the world if they can be awakened. Their sleep has been induced by endless sports events and by glorification of sensuous trivia. "Keep them dumb and keep them happy," has been the style of governments since the time of Cyrus the Great. The public is generally long-suffering but can be extremely volatile when it is aroused, and it is finally beginning to realize that global warming effects are difficult to ignore. It is a serious and possibly incurable problem to devise ways of healing planetary damage caused by human "greedocracies," whether they are labeled democracies, aristocracies, oligarchies, or whatever else keeps the people unaware of grand theft by the conscienceless elite. Earth simply cannot or will not tolerate much more human intervention.

Today's focus on hedonism has moved us to a point justifying the quip that "we are standing with one foot in the grave and the other on a banana peel." With the military toys that technological eroticism has contrived, we could be precipitated into a near global war that would wipe out hundreds of millions of people and leave the remainder proceeding toward slow death by radiation, and the plagues that follow chaotic events. The curtailment of our irresponsibility toward our Earth can only be eradicated through an awakened change of heart.

From Small Beginnings

Since this principle of the Manifesto is about the need for us to refrain from exterminating and plundering the planet, you might wonder how we have applied this to our lives. To begin with, Linda and I decided in the 1970s that we should not travel excessively, and should do as little driving as possible. We tried to live within walking distance of work, and as a rule did not start a vehicle during the workweek.

Some years ago we purchased a small acreage that had been logged and mainly clear-cut. We rearranged logs discarded by the loggers and placed them across slopes to retard erosion and evaporation. We removed eight logjams that blocked the small creek on the land and restored the fall run of kokanee (small salmonids) that spawn in such creeks. After our work, Fisheries tallied four thousand kokanee spawning in that stream one autumn. Later we were able to work with others to build a log fish ladder to help fish through a culvert that had formed a plunge pool from which they could not ascend. With the fish ladder, fish could go farther upstream to spawn. We also planted eight thousand trees, many of which we dug from roadside ditches where they are cut down every year or two by highways maintenance machines. In our spare time we collected and dried hundreds of tree cones, shook out their seeds and broadcast thousands of them on the land. Each year we raise some tree seedlings, many of which we give away to people who will plant them. We have the habit of planting trees where they appear to be needed whenever we can, and many people could do this sort of thing. Whatever we have done is too little, but I firmly believe that Earth needs our help in covering bared, often compacted land, with whatever will grow, indigenous things as much as possible. I might mention that we grow a garden to supplement our table and we eat very little meat.

In general, we feel we have no right to be heavy consumers. Life is interesting without all the paraphernalia of our technological society, and we feel blessed to enjoy the sights and sounds of the natural world. While working for hours on the clear cut, we often fell into a meditative state as often happens during manual labor. By listening to our inner voice we felt that what we were doing was right, and that it should be done. In such cases it seems that labor becomes a prayer, a form of piety. We feel that life is grace and that we have been beneficiaries of an invitation to live on this beautiful planet, which is the possession of itself and of some greater power. We sense and appreciate but cannot describe or understand these things except within certain limits of our own being. I find nothing wrong with stepping outside in the morning and saying good morning to the birds, trees, sky, to squirrels or whatever animals we might see. I am not concerned whether this would be tagged anthropomorphism, paganism, or irrelevancy. The shoe fits, so I wear it.

Shakespeare suggested, "Act well your part. Therein the honor lies." Is this matter not one of the things that should be part of the educational quest and discovery in all schools? I recall reading a story in grade school that provoked a lot of thought and discussion. It was a tale of a king with an elegantly beautiful daughter whom he wished to give away to a suitor who could answer the question nearest the king's own heart. The penalty for failure to answer the question to the king's satisfaction was death. The question required the latest (in a long line) of suitors to tell exactly how much the king was worth. The pattern of events was that each suitor would be wined and dined and would meet the fair lady. Following the custom, he would spend the night as a guest in the castle and in the morning would express his estimate of the king's wealth. Failure to give the proper answer resulted in immediate execution. Our hero and winner of the fair maiden was one who, no doubt after a night of calm repose, told the king that since Christ was worth only thirty pieces of silver paid to Judas, then he, the king, could not be worth more than twenty-nine pieces of silver. This pleased the king and the knight-errant won the fair lady. From that time to this we base a person's worth on the estimate of his or her wealth.

A Return to Idealism

Along with the general problem of over-expectation, which results from our apparent psychological need to feel important, there is the problem that leadership is not handled well by those who attain it. Pre-election platitudes vanish into thin air and new incumbents trot dutifully behind the political machine. This is probably why the statement that "power corrupts and absolute power corrupts absolutely" came into being. General acceptance of the foregoing observation indicates that there is another subject that needs to be pursued in education, which has to do with how we may train individuals to become effective leaders. Insofar as

I can see, this necessitates teaching that concepts of humility and integrity have general applicability to all students, and particularly to ones who choose forms of life devoted to a high order of public service. It is rather serious that we seem to have decided that idealism is a bit of a nuisance in this very practical world. Society has been trained to accept the idea that materialistic acquisition is the nature of the human quest. However, the fascination with voluntary simplicity that has arisen in society is a substantial rebuttal of possessions as the goal of life. Adherents refute the current view of progress and, whether or not religious, sense the practical wisdom in the observation in Matthew (15:14) "If the blind lead the blind, both shall fall into the ditch." No question about that!

If ever there was a time for heads-up thinking on the part of the nobodies, it is now. But how do those of us who are merely a social insurance number and a pictured driver's license have any clout with the "somebodies" who look upon us with a plethora of disdain?

There is an answer from Nature, which was expressed poetically in Albert Schweitzer's volume *The Decay and the Restoration of Civilization.* He expressed the thought in this way: "When in the spring the withered grey of the pasture gives place to green, this is due to the millions of young shoots which sprout up freshly from the old roots. In like manner the revival of thought which is essential for our time can only come through a transformation of the opinions and ideals of the many brought about by individual and universal reflection about the meaning of life and of the world."[17] This is why the greening of individual lives is necessary. Schweitzer goes on to say that we must recognize the sort of folly in which we have lived. This basically consists of wanting too much, expecting too much, and of failing to recognize how utterly foolish we have all become. Just the recognition of this and the desire to no longer be saddled with such grandiose expectations will serve to begin dislodging the current ideals born of vanity and passion. We can, ourselves, make the further true observation that any field that has become beautifully green becomes that way only because the individual blades of grass have caused the greenness to emerge. Since we are not very important to the movers and shakers of the world we can at least green our blades, as grass should do.

And in general, if government should (of course it should) divorce itself from the money tribe, and realize that gasoline rationing is essential and that airplanes in flight should be reduced to less than 10 percent of the present number, we can say, "hooray!" and run our machines even *less* than we are allowed. I remember an elderly couple that wanted to visit one winter. We suggested that they use an airline to make their visit, and the lady replied, "Oh Bob, flying isn't natural. God didn't give us wings." Nothing wrong with her thinking!

How can any of us deny that the quality of the seed affects the quality of the harvest? Is not the unfolding order of tomorrow, which is plainly visible, a solid recommendation that we have no choice but to move to lives of Spartan simplicity, and do so with haste?

The need for a significant change in our aspirations and consumption habits is increasingly apparent. Moral reprehensibility is involved in all those acts that exhibit gluttony. To strive purposefully to have a bigger house, newer car, more expensive clothing, and other symbols of status is evidence of succumbing to the unreflective attitude that drives our society to cut the ground out from under its feet. These acts unintentionally make us plunderers and exterminators. And this has become literally true in the area where I live as each year, in the fall when bears are normally stoking up food for hibernation, they come into towns because of degradation of their own habitats by roads, logging, and development. Looking there for fruit trees or other food sources, they are shot and killed. There is an article in a recent local paper about a man who shot a grizzly mother and two cubs that came up on his deck and began eating garbage and dog food that he stored there. Bears are also drawn to barbecues, which attract them because of the smell of food. We live quite far away from town but recognizing the bears' need, we often leave the fruit on our own trees for them and watch them hungrily making a meal.

There is something ominous in the deliberate and highly intensive obliteration of the world's resources. For example, in Brazil one acre of forest is cut or burned every nine seconds. In Canada this value is one every twelve seconds. As a result, forest species are becoming extinct. We noticed that neither loggers nor forest industry representatives seemed particularly concerned about the fate of the spotted owl due to excess deforestation. One could see bumper stickers advocating, "If you see a spotted owl, shoot it." Those bumper stickers confirm unilateral homocentricity, the widespread conviction that we are the only species that matters. The extent of human concern is to be assured that there are remnant patches of land that have been given status as wetland reserves, national and regional parks, and are thus protected. But, when we fancy that we need them for other purposes…what then?

Our scientific texts display full understanding that other species are subject to limitations in number due to carrying capacity of Earth and local habitats. Many convey the importance of biodiversity and of the intricate interrelationships between species, which confirm that a healthy world must provide space and living conditions to assure the continuance of as many species as possible. The story of Hans Brinker with his finger plugging a hole in a Dutch dyke until repairs could be made is a single reminder of the knowledge that a small hole can enlarge into a monumental flood just as a small leak can sink a big ship. Having enjoyed the pleasures of comfort and mobility, we are reluctant to change, and feel that our expectations are inherent rights, even though evidence piles up that we are approaching the elastic and plastic limits of tenure as a species. We never expected it, but we have become an unaffordable luxury on Planet Earth. If we are not merely giving lip service to love of children, grandchildren, and possibly hundreds or thousands of future generations, we will have to force ourselves out of the state of denial in which we live.

The Manifesto urges that the eternal growth ideology of the market must be renounced along with the policies and practices that result from it. Many people could help curtail the idea of growth for the sake of growth by using their excess income to pay off their mortgages, credit cards, and whatever other debts they have. In these uncertain times it would be wise to remove the burden of debt and gain the increased security that comes from becoming debt free. It is really necessary to resist the recommendations of advertisers who urge us to spend the equity in our homes to buy more manufactured trash. The advertisers want profit now and do not care a bit if you lose your home because you yielded to temptation. The market is not the friend of people, but is quite the opposite.

The Need to End Subsidies

The Manifesto also recommends the end of public subsidies to industries that pollute land, air, and water and/or destroy organisms. In his book *Perverse Subsidies,* author Norman Myers adds the direct payments made to corporations by the US government to the wider costs society is obliged to carry and comes up with the astonishing figure of $2.6 trillion. “This is roughly five times as much as the profits they were making at the time his book was written.” In 2005 the Bush administration handed a further $2.9 billion to the coal industry and $1.5 billion to the oil and gas industries.[18] The Canadian government also seems to be in pathological servitude to big business. Substantial evidence for such servitude was given by CBC’s

Michael Enright during his January 28, 2007 CBC Radio Sunday morning broadcast. Reminding listeners of NDP leader David Lewis and his 1972 reference to "corporate welfare bums," Enright went on to provide information from "On the Dole," a report provided two weeks earlier by the Canadian Taxpayer's Federation. Between 1982 and 2005, $18.4 billion in the form of 47,960 grants, loans, and subsidies were given to businesses by the federal government of Canada. Less than 19 percent of the loans have ever been recovered. Regular "gifts" to the fossil fuel industry in Canada, as Linda McQuaig mentions in her book, *It's the Crude, Dude,* amount to $5.9 billion annually.[19] In addition to such generous support to "free enterprise" by the federal government, there are huge sums given to business interests annually by provincial governments.

Subsidies to private businesses should have ceased long ago. Voters must pay attention and energetically work to remove support for such handouts. It is no wonder that many are convinced there is collusion between political and business organizations. No effort is made by government to explain what appears as at least brazen irresponsibility in giving away public funds, especially as almost half of these funds were paid to seventy-five of Canada's most profitable companies.

It is notable that industries that call themselves free enterprise always have their hands out in an effort to get them into the public purse. If they were longer on principle and shorter on greed they would pay their own way without need for limited liability, or for the rights of a person without any of the responsibilities. It is also obvious that most corporations are not concerned about their employees. They eliminate workers as fast as they find an automated mechanism to dispose of them, or out-source their jobs whenever possible. Obviously the Luddites of long ago were prescient in their realization that machines would be used to replace employees.

Problems of the sort expressed above are problems that will be in the forefront of every young person's life. Unless students receive education that poses alternatives to the current status quo, they could easily find themselves enslaved to the exploitive industries that have infiltrated legislative bodies and re-fabricated the law until they have become a law unto themselves.

We Live by Holding Hands

As a philosophy of life we need to develop a more respectful form of symbiosis with our Earth. There is a form of symbiosis called mutualism, which demonstrates that cooperation is more effectual as a lifestyle than competition. Algae and fungi cooperate to form lichens, one of the most abundant group of plants in the world. The algae can photosynthesize to make food. The fungi cannot do so, but they earn their keep by providing anchorage for the algae and by absorbing water. Their mutualistic relationship has enabled their survival through geologic ages. Another mutualistic relationship exists between many tropical plants and the birds and mammals that transport their seed. The animals receive food for their work and the plants have their seeds borne to other sites. Note also that the seeds are fertilized by the animal in whose excretion they appear. There is another successful mutualistic relationship between termites and single-celled protoza of the flagellate order Hypermasttgida. The protozoa digest the wood fibers that the termites masticate and swallow, but cannot digest for themselves. The two organisms cooperate to gain their food and neither can live without the other.

We humans have mutualistic microbes in our small intestines, which synthesize Vitamin B12, providing us with an "intrinsic factor" (an enzyme in gastric juice that enhances the uptake of B12) vital to our digestion. Sometimes an individual has an inherited inability to produce this factor, or may lack the proper bacteria following a course of antibiotics. Antibiotic, as a word, names a substance that is against (anti) life (bio). Lacking this "intrinsic factor" a person can suffer from malnutrition or may experience blood-clotting difficulty. Intrinsic-factor manufacture decreases from stomach surgery or bacterial excess in the gastrointestinal tract. Autoimmune reactions may also destroy the intrinsic factor.

People who get into the woods often see Old Man's Beard, dry-looking, usually light green lichen that streams downwards from a tree branch. Lichens of various sorts, particularly foliose lichens on old-growth trees, provide food for woodland caribou. In the north, lichens called Cladonia, also known as pixie cups, are a major food source for barren land caribou. These widespread lichens are all members of a large group of plants that contain algal cells supported within strands of fungus. Either these types of lichens or those which form atop widespread boulders indicate the success that mutualism has afforded many species.

With sufficient maturity, mankind's economic aspirations would not persist in reducing Nature to the status of a vassal. This is an unworthy aim, which displays arrogance not befitting a species that calls itself *Homo sapiens.* A nobler goal would be to live harmoniously with Nature.

As likely mutations from chimpanzee ancestors, it does not befit us to make assumptions about our superiority to other organisms. It is not sensible to claim that one species is the special product of a creator, and that other species are inferior and can be exploited at will. It certainly does not befit a devout species or a good visitor, to trash the home of its host. Possibly those individuals may be right who suggest that by having lost our instincts we must necessarily fade away before we can do much more damage to the planet. Perhaps that's the only way to get rid of corporations.

From ancient times people have acknowledged the Earth as a parent. The gender of the parent Earth seems to have varied from nation to nation. Some speak today of Mother Earth and many people have heard, no doubt, of Mother Russia or of Germany spoken of as the Fatherland. Not only has there been great love for the nation involved, but also as history verifies, there have been strong mystical bonds between people and the Earth. Actually the Native people of North America sensed the presence of a Great Spirit, an ever-present entity that permeated all things. That purification and prayer preceded and were part of a successful hunt and of the procurement of food was a form of life-consciousness that is sadly lacking today.

It might be worth reflecting on the thought that humility has long been considered an important virtue.

[1] *The Vancouver Sun,* Saturday, June 10, 1999, B5
[2] Ibid.
[3] *The CCPA Monitor,* February 2007, p.40.
[4] Anthony de Mello, *Awareness the Way to Love* (New York: Quality Paperback Book Club, 2006) p.146.
[5] *ecologist,* December 2006, p.31.
[6] *State of the World 2002,* p.14.
[7] *The Hillhurst–Sunnyside Voice,* Calgary: January 1, 2006, p.9.2.
[8] *Quantum Questions,* Edited by Ken Wilber (Boston and London: Shambhala, 1985) p.97.
[9] *Quantum Questions,* pp.122–125.
[10] Ibid, pp.184–193.
[11] Fritjof Capra, *The Web of Life* (New York: Anchor Books, A Division of Random House, 1997) p.12.
[12] *State of the World 2005,* p.123.
[13] Arnold Toynbee, *Mankind and Mother Earth: A Narrative History of the World* (New York & London: Oxford University Press, 1976) p.18
[14] Lao Tzu, *The Way of Life* (New York & Toronto: The New American Library, 1955) p.81.
[15] "ecologist Digest," *ecologist,* vol. 12, no. 3, (1982), p.1.
[16] S.J. Gould, "This view of life, the golden rule – a proper scale for our environmental crisis" Natural History, September 1990, pp. 24–30.
[17] Albert Schweitzer, *The Decay and the Restoration of Civilization* (London: Unwin Books, 1967) p.88.
[18] George Monbiot, *Heat: How to Stop the Planet from Burning* (Canada: Doubleday, 2006) p.55.
[19] Linda McQuaig, *It's the Crude, Dude* (Canada: Random House of Canada Ltd., 2005) p.326.

10: Fourth Action Principle Promote Ecocentric Governance

As author Father Thomas Merton suggested some years ago, humans have a desperate need to recover "spontaneity and depth." Both qualities, he contended, have been virtually obliterated by technological skill that has made us "rigid, artificial, and spiritually void." He recognized that tough discipline and a return to severe traditional mores have a role to play in moving us closer to the truth that every aspect of life needs to be viewed in the context of the sacred. With regard to the word "sacred," I refer to the definition in Webster's dictionary: "worthy of respect or reverence." If our general behavior changed so that we began to express reverence or respect for life, we would easily be able to begin problem solving rather than continuing to be the cause of even greater perils to our existence.

Part of the motive that led to production of "A Manifesto for Earth" is revealed in the preamble to the principles, which states, "Humanity's 10,000-

year-old experiment in mode-of-living at the expense of Nature, culminating in economic globalization, is failing. A primary reason is that we have placed the importance of our species above all else." This is an accurate statement regarding human self-centeredness. This attitude still prevails and makes the achievement of ecocentric governance a problem of great magnitude. Governance has been significantly corrupt for centuries, and is so satisfying to the minority controlling society that it has blinded them to the destruction that their arrogant mismanagement is bringing to the entire planet.

There was a time when humankind with a few simple tools was no threat to the planet. In the area where I live there are still some rock paintings left by indigenous tribes, and some local people enthusiastically collect arrowheads and other artifacts. Upper and Lower Arrow Lakes were named after the evidence remaining from early occupants. On the west side of Upper Arrow Lake, now demeaned to reservoir status, there are still signs of a largely vanished community once called Arrowhead, where a stern-wheel steamer regularly stopped years ago. Such were the days when people melted back into the Earth, neither pickled nor cosmeticized by necromancer's arts.

Gulliver's Travels

It does seem that there has been little improvement in the quality of government since the days of Jonathan Swift, a priest in holy orders who was appointed dean of St. Patrick's, Dublin, in 1713. In his satirical fantasy *Gulliver's Travels,* Swift focused on the organization and behavior of states, and thereby created a masterpiece that has become a classic. *Gulliver's Travels* emphasizes the mischievousness of the human mind and probably reflects the author's religious viewpoint that man's sinfulness can be altered only by divine intervention.

Gulliver's Travels relates the visits of Lemuel Gulliver, an English marine surgeon, to four imaginary countries. The fourth country he visits is the land of the Houyhnhnms, a place where the dominant species is composed of virtuous horses who are reasoning beings. In the land of these intelligent horses there are large numbers of filthy, uncouth, vicious, untrustworthy brutes known as Yahoos. These were basically uncivilized humans lacking reason. Gulliver nearly falls into their hands but is rescued by the horses that are troubled by his close resemblance to the Yahoos, except that he is clothed and displays reason. He spends considerable time in conversation with a master horse that quizzes him about his place of origin and the behavior of the slightly different group of Yahoos among whom Gulliver lives. The master horse is a peaceful and kindly being, unaware of the concept of war. He is shocked to hear of the disreputable behavior of Gulliver's compatriots.

Gulliver described war and told the master horse of the long conflict with France conducted by the Prince of Orange. "I computed at his request, that about a million of Yahoos might have been killed in the whole progress of it; and perhaps a hundred or more cities taken, and five times as many ships burnt or sunk."

Responding to questions about motives for war, Gulliver replied, "I answered that countries went to war with one another for innumerable reasons of which I would give only a few: Sometimes the ambitions of princes who never think they have enough land or people enough to govern, sometimes the corruption of ministers, who engage their master in a war, in order to stifle or divert the clamour of the subjects against their evil administration...Neither are any wars so furious and bloody, or of so long continuance, as those occasioned by a difference of opinion, especially if it be in things indifferent...Sometimes one prince quarreleth with another, for fear the other should quarrel with him...It is a very justifiable cause of war, to invade a country after a people have been wasted by famine, destroyed by pestilence, or embroiled by factions among themselves."[1]

In reading this story we can easily see that ethics and civility in government haven't made much progress in nearly three centuries.

War in the Middle East

We now face a critical need to fortify existing governments by passing ecocentric laws, regulations, and programs that respect Nature for its wholeness rather than in the fragmentary manner now followed, which dissects Earth for its parts whenever they create profit.

Consider the war launched by the US against Iraq. The terror issue that promulgated the war was apparently falsified or at least "mystified." We now know the claims that Iraq had weapons of mass destruction were untrue and unsubstantiated. Many believe that the fear these claims created provided enough support for the USA to launch their war against Iraq. Quite a few people also fear that the hullabaloo about bringing democracy to Iraq masks serious efforts on the part of government to substitute fascism for freedom right here in North America. There is a widespread belief that control of oil supplies is probably the largest factor in the militaristic mess that has been created. The US is obviously looking for a way out of the war and the president's popularity has become pretty marginal, or "skimmishy," as an old settler in the woods, a neighbor of mine, refers to it. The Afghanistan situation is perhaps ancillary to the Iraq dilemma, and may have more to do with an oil pipeline that the Taliban refused to let US oil giant Unocal build from the Caspian Basin, south through Afghanistan to Pakistan.[2] The war seems more connected with refusal to cooperate with the petrochemical industry than with the threat of terrorism, or a desire to help the Afghanistan people.

The carbon dioxide crisis, to which humans have given birth and rapid growth, is so stupendous and so likely to be lethal to the human race that the continuation of these high tech perpetual stalemates is quite serious evidence pointing to the fact that governments are acting without ecological wisdom. The world's politicians have worked for years to downplay the serious problem of human pollution, with fossil fuels use being but the tip of an iceberg. Warfare is a banquet of fossil fuel use and of the products of the munitions industries. The planet cannot afford it; governments, which have many more important things on which money should be spent, cannot afford it; the public cannot afford it. The death sentences handed out by governments to the victims of warfare have been a display of the utter callousness of generations of politicians and profiteers who have never realized that peace on Earth would make a true civilization possible. Armies have always provided an open invitation to those who seek power at any price. There are other alternatives. One of these is to keep all problems on top of the table.

Leading by Example

What does that last sentence mean? It reminds me of the sad fact that the public is not really kept informed of forthcoming crises, nor invited to participate in situations approaching crisis level. I am reminded of a book I read, entitled *Asking for the Earth.* I liked that title and believe that changes in human behavior by stopping wars and by greatly reducing carbon output are actions needed on behalf of the planet. So this book, too, is an asking for responsible behavior of a sort rarely displayed by governments or private citizens.

In the above book, James George spoke of a dinner he had with

Maurice and Hanne Strong in Toronto. George is a retired Canadian ambassador with keen environmental interests and Maurice Strong served as Secretary General of the UNCED Conference in Rio. Strong mentioned that at the Rio conference some attendees felt that he was overplaying his hand in talking about world problems. George felt differently and stated that Strong had perhaps been underplaying his hand, not feeling that he had fully revealed the depth of the dilemma that exists. Strong explained that the full truth of the human plight could frighten people so much they might basically give up and conclude that it was too late to do anything.

I do understand Strong's attitude, but I believe that the longer those in power continue to hide evidence of the grim situation that now exists, the more people will continue self-denial and say there is nothing they can do, or else rely on the technocrats to find a way to solve the problem. Good leadership, which reveals that the leaders are actually leading from conviction, would spawn action once the public understood that its leaders could be trusted. This is a real problem to many of the ordinary individuals with whom I come in contact. They would react more favorably to leaders who rise above the "don't do as I do, but do as I say" form of governance and would like to see governance by example. In 2005 in British Columbia there was a teachers' strike focused on pay increases and a new contract, to which the government was adamantly opposed. Shortly thereafter the same provincial government announced what most people deemed an exorbitant raise in salaries and benefits for all members of the legislature. The public response was so indignant the government had to back down. This did raise a question in many people's minds. They asked how come politicians could give themselves raises or benefits? The only answer I can envisage is that a formula should be involved that relates political salary increases with average increases in a number of other occupations. Politics is not quite in the same league as a privately owned business where the boss sets his own salary. Public servants must always keep in mind the entrenched axiom, "One must avoid not only evil, but also the appearance of evil."

Throughout the world, chief executive officers and affluent people in general have been getting a larger share of income while people at the bottom of the financial heap are struggling, not always successfully, to stay alive. One would think that the affluent would be so ashamed of their own greed that they would retreat into monasteries to do some soul searching, if they have been issued such an entity. Politicians knowing about these things should be establishing "level playing fields" that have to do with more than merely a global economy. Strange isn't it, that some CEO gets a few million dollars raise for a single year while more than 852 million people worldwide remain undernourished - that's 13 percent of the population.[3]

I also recognize that it is very difficult to work for the public. It's tough to have a few million bosses. But if one aims to be an effective leader, that individual must operate from a deeper sense of justice than seems evident at present.

Society has, unfortunately, created roles for people to play. The role of politics has been one of endless compromise and knuckling the forehead to the empty-headed elite who haughtily rule the world. Now that Nature has shoved us into the corner we have been constructing, the only smart thing to do is what planetary ecology demands. In the final analysis we aren't any more important than a plague of locusts or jackrabbits. We have to act, not merely expect divine solutions. If we received a divine solution we might find all our wheels turned square some morning while our chainsaws and mega-machines might be turned into sponge rubber.

The issue Maurice Strong brought up about the public not being able to handle the truth constitutes a real problem, but the longer people are fed poppycock to make them docile, unrealistic, and needing soporifics in place of hard fact, the longer it will take to bring about realistic solutions.

We should start with absolute truth from government. Politicians must realize that they are not elected to be members of a private club that decides what things the public should be told and what things should be kept from it. In the fullest sense of support, they are public servants and no servant would be respected who kept dangerous secrets from his employer. In the issue before humanity today, millions of people are guilty of serious crimes against the Earth. People in power hate to admit that the planet has been torn apart for profit or convenience since our ancestors climbed down from trees. We have simply been biting the hand that feeds us. We should be forcing our politicians to move away from procrastination. The public needs governors who will fill their shoes completely and say, "Hey folks, we live on Planet Earth, and a mighty fine place it is, but if we can't put away our toys and get serious about protecting what keeps us alive, we are nothing but burnt toast." It's tough to have to realize this, but most people have been putting off getting down to reality. The facts have become too obvious to deny. Earth is in big trouble, and what has been considered to be progress has been the exact opposite. Whatever it is necessary to do — gas-rationing, reducing airline flights, canceling the Olympics and other spectaculars, stopping wars all over the world, replanting forests, stopping logging — we do it. Maybe in ten or twenty years a wiser generation can slowly resume some of these activities — but perhaps it will know better.

Everyone seems to be waiting for leaders to take on leadership, but they are still playing the old game of minimal compromise. When I ponder over the sort of individuals needed to lead the world to survival I am reminded of a character called Mr. Fidus Neverbend, who was created by Anthony Trollope. It was said of Neverbend that he "was an absolute dragon of honesty. His integrity was of such an all-pervading nature that he bristled with it as a porcupine does with his quills." I haven't heard of a quilled legislator, but note that British Columbia's NDP MLA Corky Evans bristled a bit in his reply to the 2007 throne speech. In essence, what Evans did was to chide the present government about its lackadaisical stand on global warming.

On a national scale consider the highly polluting Alberta tar sands project. That's one issue that should be abandoned because it is an ecological travesty. It could be put off and done more sanely, perhaps in fifty years; or we might decide by then not do it at all. At a time when damage caused by the burning of fossil fuels has become so evident, it is ludicrous indeed to be burning natural gas to extract bitumen from tar sands to make more fossil fuels. Not only that, but 359 million cubic meters of water is being taken annually from the Athabaska River for tar sands development, and this is to be increased by 50 percent. The present amount of water being used is twice the annual requirement of Calgary. This is happening at a time when energy needs should be reduced for the sake of survival. The Sage Centre and World Wildlife Fund is calling for a moratorium on the project.[4] It just does not seem at all wise to continue with a project that will produce colossal CO_2 emissions. The tar sands project is too much like a fly buzzing around someone's head until it is swatted. In this case Earth may deal the blow. It would be much wiser to show restraint and demonstrate a bit of ecocentric understanding.

Some Success Stories

Can governments do good things? Of course they can. But to do good things, governments would be better off if they focused on present problems and national solutions rather on international problems that direct our focus outward rather than inward. It has been noted that Bogota, Columbia underwent a great, favorable change during the government of Mayor Enrique Peñalosa, whose goal was to improve quality of life in that city. Prior to Penalosa's election Bogota was a site of civil war and violence. During his tenure, school enrolments increased by 34 percent or 200,000 students. One thousand two hundred and forty three parks were built or rebuilt, both small and large. One and a half million people now use these parks yearly. An effective rapid transit system was constructed. In addition the former high murder rate per capita in Bogota, was reduced to a level lower than that of Washington, DC.

Peñalosa's own vision was significantly important. At the time he became mayor a 600 million dollar elevated highway was planned. Substituted for the highway was a cheaper rapid transit system that used existing bus lanes. The system now carries 780,000 passengers daily, and 15 percent of passengers are car owners. The mayor also caused hundreds of kilometers of bike paths to be built, and set aside pedestrian-only streets. New public libraries and schools were added to the community and helped to direct society toward greater emphasis on people and community life. Peñalosa's own philosophy was expressed in his observation that "a city is successful not when it is rich but when its people are happy."

The foregoing tale appeared in a *State of the World 2004* article written by Gary Gardner and Erik Assadourian. Their comment following the Bogota success story was significant in light of today's focus on wealth as a primary goal: "By redefining prosperity to emphasize a higher quality of life rather than the mere accumulation of goods, individuals, communities, and governments can focus on delivering what most people desire." This, in effect, is the good life redefined.[6]

The same article mentions an organization in the US called The Center for a New American Dream. Its slogan may reflect people's growing realization that the old dream has become a nightmare. The Center recommends that people focus on "more fun, less stuff." Fundamental conservative behaviors are encouraged, things as simple as reducing meat consumption, reducing water heater temperatures, and switching to water-efficient faucets. The group of 14,000 members claims that its actions reduced carbon dioxide emissions by four million kilograms and saved more than 500 million liters of water. Obviously, thoughtful individuals, communities, and governments can make a big difference. I do believe that governments need to step forth and catalyze actions that would cause great numbers of people to unite in planet-saving behavior. Once people's momentum is activated and they are encouraged to put on their thinking caps, the limits suggested by Kyoto Accord targets could be far surpassed - enthusiastically surpassed.

In a major essay, "Engaging Religion in the Quest for a Sustainable World," *State of the World 2003* offers substantial evidence that religions are beginning to take an active role in working toward protection of the planet. Their potential impact is formidable. Five powerful assets at their fingertips are: their ability to affect worldviews; their moral authority; their billions of adherents; their wealth; and their proven capability to influence and affect society from community to international levels. Behind all these power assets lurks, I would say, an imposed obligation to protect the works ascribed to divinity. Sixty-five percent of the people on Earth share three major faiths: Christianity (33 percent), Islam (19.6 percent), and Hinduism (12.4 percent). What is even more impressive to me is that in forgotten ecumenism they could become a mighty force for peace on Earth - a now extremely urgent need since the ecosphere cannot afford more of the incredible thermodynamic impacts caused by modern warfare.

By using the pulpit to address the global crisis brought to a head by global warming, by using congregational newsletters, websites, their own magazines and newsletters, plus op-ed articles, by urging members to write letters, instigate or join boycotts and protests, by use of their own physical facilities for meetings and educational presentations, by revising their purchasing and investment options, by working to strengthen community ties, and by helping people to shift from material aims to spiritual ones, thereby simplifying their own lives, they could enrich all humanity with a more noble worldview. An immense effort toward peace on Earth and peaceful respect toward Earth might help bring about a worldview in which brutal selfish competition is finally subordinated by selfless cooperation. Deeper emotional ties to the life-sustaining Earth could serve as an essential factor in reuniting our severed physical-spiritual reality.

If religions put their shoulders to the wheel and taught that true dedication to the Golden Rule means far deeper respect and love for the Earth, civilization might then rise upon its feet. This does remind me of the serious rejoinder that Mahatma Gandhi made to members of Queen

Elizabeth's court when he was asked what he thought of civilization. His response was brief, "It would be nice!"

The Manifesto speaks with clarity of broad change that needs implementation. It provides a direction for public response. There is graphic evidence that while the rich get richer and more self-centered, the world's poorest move steadily toward starvation and death. The lowest one-tenth of wage earners has shockingly little income and the top tenth becomes more glutted each year. This is par for the course in our world, and shows that present governments make no serious effort to rectify problems for which determined leadership is vital. The disparity of earnings that has been formulated by the business world is a disgrace to its initiators and to governments that could devise tax structures to prohibit both cutthroat profiteering and rape of the world's ecosystems.

The Denial Machine

An emergency has been created that is obtusely ignored. For years corporations have paid willing individuals in the scientific community and lobby groups to deny global warming, and facts are obscured or altered to suit their single vision. Note this surprising reaction: "In an unprecedented step, the Royal Society, Britain's premier scientific academy, wrote to the oil giant (Exxon Mobil) demanding that the company withdraw support from dozens of groups that have 'misrepresented the science of climate change by outright denial of the evidence.' The scientists also strongly criticized the company's public statements on global warming, which they described as 'inaccurate and misleading.' The Royal Society cited its own survey, which found that Exxon Mobil last year distributed $2.9 million to 39 groups that the Society says misrepresent the findings of climatologists. This is the first time the Royal Society has written to a company to challenge its activities."[7] The Royal Society stresses that there "is overwhelming scientific evidence that greenhouse gas emissions are linked to global warming."

A problem exists in most communities regarding the suppression of information that could be described as ecological in character. Largely through manipulation, a prejudice has been created that presents environmental concern as redundant or fanciful. Dominance of the economic worldview shared by government and industry offers convincing evidence that their rigid stubbornness flies in the face of sensibility. When a major problem such as global heating reached an undeniable stage, the inflexibility of world leaders, solidified by their unreal expectations, found denial the only tool that their single-vision afforded.

It is easy to see that human values — and there are yet such things — are sabotaged by the indifference of governments and by their failure to live up to traditional ideals. Governments must support the integrity of Earth and the needs of the deprived and needy with at least as much enthusiasm as they support the subsidized, extremely high profits of industry. Though the dominant powers in society bristle at anyone using the word "rape" to describe what is pretended to be development, industry threatens the existence of many species, including our own, by perpetuating an economy that is now a rapidly metastasizing terminal cancer. The metastasis increases speed with each blatant abuse of ecosystems. The UN Millennium Report's conclusion that most of the world's ecosystems are seriously degraded is merely confirmation of what many people have seen coming for years.

The Manifesto offers a valid recommendation in stating, "Homocentric concepts of governance that encourage over-exploitation and destruction of ecosystems must be replaced by those beneficial to the survival and integrity of the Ecosphere and its components." It goes on to say that "advocates for the vital structures and functions of the ecosphere are needed to be influential members of governing bodies."

Ecologically ignorant, governments are self-important bodies absorbed in their own agendas and preferences. People, with government cooperation, have been conducting war against the planet for years. If humans knew enough to overcome their greed they would realize the foolishness of the quest to conquer Earth. In describing our assault on Earth, Aldo Leopold claimed, "We are mining the Alhambra with a shovel, and proud of our yardage." Now repeating history, there are numerous Neros fiddling in office while the world's Romes are gasping for breath amidst flatulence from vehicles and factories. Meanwhile storms are sporadically huffing and puffing preparatory to blowing down our houses. Unless our industrial society finds a quick way to educate incumbent politicians, or finds a truth serum that will awaken them to the real issues, the world's erratically guided ships of state, will beach themselves on the shoals of incompetence.

An interesting example of Canada's reluctance to do much more than procrastinate about carbon dioxide production, while denying cooperation with the Kyoto Accord, is given in George Monbiot's book *Heat* —

How to Stop the Planet from Burning. Monbiot's columns in *The Guardian* are read around the world and are celebrated for their research and depth. His book contains a special foreword to the Canadian edition. In it he acknowledges that Canada is a country of grueling winters and long distances between many places, but advises Canadians that the sustainable limits for carbon dioxide emissions amount to 1.2 tonnes per person annually and that amount is only one-sixteenth of "what you currently produce." He lambastes the Canadian government for its refusal to reduce carbon-dioxide emissions and particularly takes Prime Minister Harper and his former environment minister, Rona Ambrose, to task. He bluntly contends that the government's excuses "are an astonishing instance of political cowardice." Disclosing the servile attitude of government toward industry, he states that whereas Harper claimed to the people of Canada that he could make tough choices, he has shown himself to be "an irresolute wimp as soon as he was faced with a choice between upsetting a few industrial lobbyists and helping to save the planet." Monbiot's own figures "suggest that Canada should cut her carbon emissions by 94 percent between now and 2030." He states that in the United Kingdom individuals each produce an average of 9.5 tonnes of carbon dioxide per year. In Germany 10.2 tonnes per person are produced. The French turn out 6.8 tonnes per year. Canadians produce a whopping 19.05 tonnes per capita. The two other world leaders are Australians who produces 50 kilos more than each Canadian, and the Americans who turn out a tonne more than each of us.

To look at the matter from a different perspective, The Tyndall Centre for Climate Change Research announced that, "we need worldwide to reduce our carbon emissions by an unprecedented nine per cent a year for up to 20 years." I suggest that we memorize this figure of 9 percent a year and make it a part of our personal goals. It is especially important to explain it to our children. I will repeat again Monbiot's statement that 1.2 tonnes of carbon dioxide should be our maximum emission for the year. It definitely means scratching airline travel from any itinerary.

What about the Economy?

Realize that our economy has been satanic. That is a bad implication, but let's look for example at remuneration for chief executive officers and consider the lack of concern exhibited for the unprivileged of the world, or even for lower paid corporate employees. The following article appeared in the November 2006 issue of *The CCPA Monitor,* a publication of the Canadian Centre for Policy Alternatives under the caption, "CEO pay hits new heights in U.S.": "Average annual individual CEO compensation in the United States rose from an average of $3.7 million in 1993 to more than $10 million after 1998, reaching a high of $17.4 million in 2000 when the stock market peaked…The database is comprised of 1,500 companies that account for more than 80% of business firms in the U.S. in terms of market capitalization. The definition of executive pay in the study includes cash, bonuses, and equity-based pay such as stock options. The bottom line of the study is that the combined pay of the top five executives of these companies approximately doubled over the 1-year period studied, from 5% to 10% of company profits. In total, the top five executives of these 1,500 companies (7,500 of them in all) were paid $350 billion over the 10 year period, and currently are paid a total of $40 billion a year. The authors find that almost none of the big executive pay increases can be explained by their firms' improved financial performance. Their pay packages were enlarged regardless of whether their companies' profits rose or fell."[8]

There is an opposite approach to wage distribution that is very interesting. This was evidenced by the Scott Bader Co. Ltd., a more idealistic business formed in England by Ernest Bader in 1920. Although the company led a quiet revolution against standard business practices, it grew from 161 to 379 employees and succeeded in demonstrating a more humane approach to economic methods. An outstanding part of Scott Bader's business approach was that the lowest and highest paid workers differed in remuneration in a one-seventh pay scale with the highest paid worker receiving no more than seven times the salary of the lowest paid worker. If you wish to peruse this further, see pages 272–292 in E.F. Schumacher's *Small is Beautiful.*

Modern corporations and the English Scott Bader Company represent two different kinds of thinking. We seem to have moved today toward some sort of mockery of the entitlement of various sorts of individuals. Is it not strange that the forces that have recognized the need for a minimum wage to protect the poor have not seen the even greater need for a maximum wage, if for no other reason than to protect society from the rapidly escalating insanity of salaries beyond any conceivable need?

Beginning the Battle

So, where does one start? Obviously at home and locally. This involves real community newspapers with editors that seek to awaken readers. It involves individually reducing the use of gas-guzzlers of all shapes and descriptions. It involves politicians paying attention to their own people and own countries. They probably do not know it, but the bottom line truth is that humankind is now engaged in a battle to prevent sinking into chaos, wars, famine, disease, and exhaustion of resources.

Look around yourself. What needs saving in your own community? Is there good arable land nearby? Why are we not encouraging local, organic farming? Why are local watersheds being destroyed? Why are we not composting local waste and perhaps even generating local biogas resources from our own sewage? Why are we not waking up politicians to encouraging local sustainability? Why are we not seeking increased local governance? Why are we not learning more about taking care of our local region?

Realizing that the ecosphere sustains our existence, it is shocking to observe the fantasy world that surrounds political thought. It is apparent that corporate propaganda has deluded politicians and exposed their subservience. Prime Minister Harper lost his chance to move into a position of real leadership. The bombastic rhetoric of Canada's environmental leadership is no more significant than a pebble rattling around in an empty can — noisy but meaningless. If our prime minister had committed Canada to double the obligations it would have as a signatory of the Kyoto accord, rather than suggesting with his Clean Air Act that we would come around to it somewhere before 2050, he would have displayed the tough leadership about which he bragged. Then he might have become an admirable model for all other nations.

The Manifesto authors speak of an imperative need for "eco-politicians knowledgeable about the processes of Earth and about human ecology" who will "give voice to the voiceless." These individuals will speak on behalf of animals being pushed toward extinction and reveal "the necessity of legally safeguarding the many essential non-human components of the Ecosphere." A step forward will occur when it is recognized that other organisms are Beings, not just things for our use or amusement.

Pretending that environmental problems could be obscured by sidetracking them to "environmental hearings," authorities have deferred ecological problems, obscuring them in oratorical whimsy and consistently underestimating natural forces. The idea that many "developments" are actually acts of vandalism against ecosystems has not penetrated the homocentric devotion of legislators and judges knowledgeable about human law but untutored in natural law. A great shortcoming of society underlies copious legal data that sustains exploitation but lacks even rudimentary legal material designed to protect Nature. Laws in general have been devised by lawyers who display elegant verbosity to defend human rights while knowing nothing about basic life-sustaining ecological conditions. Nature has been left "behind the eight ball."

The Manifesto authors speak of the necessity of laws that are designed to protect the Earth and to move people toward ecocentric behavior. They also refer to a shocking truth, that neither Earth, nor wild animals of any kind, has any standing in court. Humans as the dominant animal species have said by their actions and laws that other animals are merely tolerated. If they become nuisances, and it takes very little for an animal to be described in that manner, they are simply killed. The Manifesto asks, "Who speaks for wolf?" and "Who speaks for temperate rainforests?" The answer is that nobody speaks for these things, and the industrial forces of the world seem bent on recklessly destroying all sorts of species habitats as well as the larger ecosystems that support them. The visionary scholar George Perkins Marsh warned in 1864, "that humans had already left parts of Asia Minor, Africa, Greece, and Alpine Europe as desolate as the moon." Protesting the lack of protection for Earth by the courts, he commented that Earth undoubtedly has standing in a higher court than any of our own. He cautioned long ago that continued irresponsible behavior on the part of humans could easily lead to the "depravation, barbarism, and perhaps even extinction" of our species. It seems that that our species can muster only minor concern for Nature and is too focused on immediate wants to develop any comprehension of a more noble universal concern.

Heaven Is Under Our Feet

As an example of the lack of concern for ecological stability, George Bush, certainly not a materially impoverished person, is part of a multinational mining endeavor intent on mining a large deposit of gold, silver, and other metals. The deposit is located under two extensive glaciers in the Andean-Cordilleran region on the Argentine, Chilean frontier. The glaciers are the source of water for a large farming region and this water supply will be destroyed by breaking apart the glaciers and creating two huge tunnels, one for mining ores and the other for storing tailings and other debris. It is apparently inconsequential to these developers that the rivers flowing from the glaciers will be "killed" by cyanide and sulfuric acid used in the extraction process. The rivers will become lethal for animals, including humans. The poisoning will last for generations. The Chilean government, apparently unwilling to govern by protecting its own terrain, has issued the proper permits and the region's powerless farmers are vainly trying to mobilize international opinion to protect the country. Does it not seem unbalanced that supposedly intelligent people will destroy ecosystems to remove gold from the ground in order to hide it in some secret vault? It may be worth hundreds of thousands of dollars an ounce the way our society is going, but a ton of gold doesn't have the nutritional value of a single apple, a head of lettuce or a potato. It is notable that pack rats are very attracted to glittering objects, which they store in their nests. The human relation to pack rats seems verified by the unreflective ambitions of those who would destroy productive land in quest of so-called precious metals.

We have created such a mess on the planet that our only hope of continued existence is to work with might and main to stop our abuse. The most important thing in our world is Earth. A new emergency economy that should replace the jaded, present economy is one that will recognize that Earth should be our first priority — always. Once again, Schumacher's thought: "Guard the health, beauty, and permanence of land, water, and air, and productivity will look after itself."

Ecocentric Thought

I suspect that it is prophetic, and not merely prejudice, to say that we are coming to the end of life's trail if we do not make a major shift from people's priorities to a more enlightened ecocentric focus. This will come about only as a renewal of thought. We must not delay in enacting laws protecting biodiversity, and we are overdue in conferring "legal standing on vital structures" for protection of their uniqueness and their contribution to ecological stability. Ecocentric thought is a step upward to a form of education we have been denied, and this sort of outlook must become a functioning part of government at all levels: village, city, county, regional district, province and nation.

The people of British Columbia are becoming very concerned about severe snow storms and unusually bad weather, such as the heavy snowfalls, extreme winds, glare-ice and blown over trees that plagued the Vancouver area in the latter part of November 2006 and did an estimated nine million dollars worth of damage to Stanley Park. Global warming is becoming a palpable reality and people now recognize that the longer ameliorating change is delayed, the more serious problems will become. The desire for change and the need for sober thought suggest that the social climate is now more receptive to ecocentric concepts.

Political candidacy at any level, from village and urban to provincial and federal, may be an option for some individuals, and I would venture that an individual with great ecological concern might gain a larger following than would have been attained a decade ago. I remember talking with the late Richard C. Passmore (Dick), the executive director of Canadian Wildlife Federation more than three decades ago, when he and I traveled together to give lectures on ecological topics. It was his view that mankind was aware of the first three dimensions: length, breadth, and depth; but was unaware that we are in a race with the fourth dimension: time. Our ecological transgressions, stimulated by our economic demands, were even then becoming too severe for our planet. We discussed whether or not humans were too dense or too reckless to realize that all species are weighed in the natural balances, and the species that are found wanting become obliterated. I was not surprised when I lately came across a copy of Toynbee's last volume, *Mankind and Mother Earth,* to read his statement, "Man, the child of Mother Earth, would not be able to survive the crime of matricide if he were to commit it. The penalty for this would be self-annihilation."[9]

The third word in the Manifesto's Principle 10 is governance. This quality of being must begin with self-governance. Moderation is probably the key word affecting most of the actions that should be taken in our present hours of crisis. The constant pressure for increased consumption that is exerted by media advertisements is truly counterproductive. The United States and Canada, with 5.2 percent of the world's population, are responsible for 31.5 percent of the world's consumption.[10] As stated in the Manifesto, "Homocentric concepts of governance that encourage over-exploitation of and destruction of Earth's ecosystems must be replaced by those beneficial to the survival and integrity of the Ecosphere and its components."

Obviously, if we were to promote ecocentric government, persons elected by the people as politician-governors would have to be versed in enough subject areas to understand more about the functioning of Earth than they know at present. It seems logical that if individuals must attend military academies to prepare them for governance of troops, there should be educational institutions that provide training for those who would govern provinces and nations. It is also obvious that basic science training in ecology should be a strong component of such training, and that formal training in ethics should also be included. At present there are no educational requirements to hold political office. There is also very poor behavior in legislatures, which speaks ill of the quality of persons who are elected to positions of such trust. I remember an elementary school teacher I knew, who had visited the federal and many provincial legislative sessions in several places in Canada. She told me that she required far better behavior in her classrooms than she saw in legislatures. Certainly more knowledge and decorum than now exists among those who govern must become part of the awareness of those who are capable of realizing that government is serious business.

Education Begins at Home

Governance is such a serious business that self-governance should be learned from childhood onward. Self-governance obviously should be learned at home and has been made more difficult for parents because of the easier recourse of plopping children in front of television. I suspect that the rapidity with which stimuli are introduced to sustain children's interest is a real no-win solution. When they enter school and need to employ concentration they are stymied by the fact that they have been trained to expect continuous entertainment. The most successful students in schools are those who have learned to behave, and to respect the rights of others. Several lessons that can be learned at home are that rights are derived from responsibility and responsibility sometimes curbs rights; that liberty is a right but should not turn to license; and that "I didn't think," and "I forgot" may be statements but are not valid excuses. "Please," "thank you," and "pardon me" are also social expressions that need to be learned by many.

Education today is unfortunately shaped to a pattern that offers all the knowledge and understanding that is desired in a consumer society wherein addiction to machinery and gadgetry is a poor substitute for a thinking public. If we want to consider what education might and should be, we should study the "liberal education" that was standard for capable students in Hellenistic culture. At the age of seven, students began learning the basics of reading and writing, undertook gymnastics, which consisted of physical training rather than merely games, and soon after advanced to studies in poetry, music, literature, and mathematics. Memorization and recitation were standard educational techniques. Students learned vast amounts of Greek poetry by heart; this included Homer. Discipline was exceedingly strict. By the time they became teenagers, they engaged in a broader and more intense study of all the subjects encountered in younger years. History, philosophy, and theoretical science were added and yet more poetry was learned. Rhetoric or public speaking was a main feature of education in its later stage. Students learned this art by arguing and discussing historical or impromptu situations. The intent of the educational process was that students would be able to function capably in civic society. At that time education was for boys only, but education of this sort should be mandatory for everyone in society today. As for computers, television, and other devices that effectively distract from what could be learned, time spent on these things merely impede good, basic learning, and their real purpose has to do with support of industry, and with indoctrination designed to make young people totally dependent on mechanical devices. I appreciate the benefits I have had from a stricter educational format that was in place in my own school years. We were brought up with the expectation that it was our job to learn, and we expected discipline in schools because discipline existed at home. Good discipline is a benefit and sets the proper environment for learning. Youthful chaos leads to adult chaos, frequently visible today.

Needed: A New Set of Ethics

The blanket of carbon dioxide we have exhausted into the atmosphere threatens us with heat such as we have never envisioned, with miserable lives of discomfort, and with extinction. The alternative to being well-cooked and starved to death is to realize that we are experiencing a moment of truth, and our survival depends on giving up many of our habits. We can reduce use of atmospheric-polluting fossil fuels by abstaining from airline travel, and by reducing our automobile use, by using railroads to move cargo when long distances are involved, and by eliminating such industrial excesses as chlorine-based pulp production and slash burning, which in one year alone in British Columbia (1988) emitted 34.1 million tons of carbon dioxide into the atmosphere (more than a billion tons in thirty years,) and by dozens of other techniques.

Employers in industrial service stress doing things only in the company way. In boot camp the fledgling service person soon hears that there is "a right way, a wrong way, and the military way." He is also advised that his job is to do things the military way, which is to follow orders without question. Throughout history there have been extremely grave abuses committed in the name of following orders. The Nazi excuse *befehl ist befehl* (orders are orders) was found to be an unacceptable defense by the war crime tribunals. But one sees even in peacetime that the principal ethic expected by an employer is that the employee be merely an echo of the propaganda put forth by the company. A private or personal ethic is frowned upon and companies themselves very often give only the slightest lip service to ethical behavior. Thus there is a long-standing history of unresolved ethical abuses, such as Union Carbide's lethal release of toxic gas in Bhopal, or the Exxon Valdez oil spill travesty in Alaska. The sum assessed against the oil company was unpaid for years, and finally reduced hugely to match the amount covered by the company insurance. Courts thus make a mockery of corporate responsibility. Under the quite deleterious-to-humanity camouflage of corporate limited liability there have been many abuses of justice. Today's double standard in law is nothing less than a continuation of the hundreds of laws passed in Great Britain as enclosure acts, which sacrificed a peasantry to the whims of assorted dukes, earls, and lords. It is a trifle amusing that blue blood is venous blood filled with carbon dioxide as opposed to red blood, which is freshly charged with oxygen and vitality. At any rate, the euphemism of the time of the enclosure acts was that, "The law locks up the thief who steals the goose from off the common, but leaves at large the larger thief who steals the common from the goose."

The self-destructiveness of society attests to the fact that the only alternative to total collapse that we have is to shift to an ecocentric philosophy of life, moving our own priorities downward and ecological priorities upward to central focus. We would have to become serious and humble, though we are not inclined that way. We would have to study how to become devoted stewards of our home planet, Earth. We have been conditioned to treating Earth as simply a place to live; we have never looked around and taken time to realize that we can be erased as simply as a pencil mark can be erased from paper.

An ecocentric outlook would bring about such a state of maturation that even prime ministers and presidents would realize that each new undertaking should be prefaced with the question "How will this affect the Earth?" For the first time in the soaked-with-gore history of human behavior, understanding would grow that warfare is totally redundant. The primary victim of all wars has been the Earth. We constantly whittle away at Earth stability in our bellicose idiocy. In spite of gorgeous uniforms and clusters of medals, in spite of bands playing, flags waving, and politicians exhorting patriotism, warfare is a sign of ignorant disrespect for the gift of life and the world we should protect at all times. It is no conundrum that the present dangerous state of affairs will soon come to a head.

So we need more ecocentric governance, more concern for land, more protection for the planet and for individual species. Where do we begin? How do we get a new consciousness to replace the ruts in which governance seems entrapped? I have a theory about that. Present provincial governments and the federal government are trapped in the front row pews they share with big business. Both forms of government pay as little attention as possible to society's peons. It seems logical therefore that changes in governance start from the ground up. This means from local communities and, even more fundamentally, from local voters. It is not atypical for dominant businesses to get their supporters elected by village or city councils. It has not been customary, though, for persons concerned about planetary health and protection to either nominate or support candidates who would understand and support ecocentric governance. This would be a basic necessity — to get recognition and influence local representation. Once representation exists, an effort should be made to develop ecocentric possibilities that the local area affords.

In the area of Nakusp, British Columbia, where I live, there is a large amount of arable soil that could be farmed and thereby provide fresh tree fruits and vegetables including potatoes, which are indeed heavyweight crops when one considers the cost of transporting them long distances. Note also that transportation distances and costs could be reduced by locally produced foods; and fresher, less-polluted foods could result from organic farming. In addition, consider that under extreme climate conditions that involve road blockages of one sort or another, local foods are of immense strategic value compared to ones that must be imported. Communities could create jobs and keep money in the area by subsidizing local producers through tax breaks, educational, or other incentives. Local canning and freezer facilities might result from the foregoing suggestions and a community might grow enough produce to gain support from nearby communities.

One small community, Leaf Rapids, Manitoba, banned retail stores from using plastic bags for goods. The community spent considerable sums for cleaning up litter of plastic shopping bags and felt it would save money and messy garbage by supplying people with free canvas shopping bags. These were passed out at a community inauguration of the new legislation. Leaf Rapids has a population of 600 people, but it might be noted that San Francisco, Rwanda, and Bangladesh have also undertaken a similar action regarding plastic bags. Thinking about the fact that almost the main use for petrochemicals is plastic products, it would seem that many communities should hop on that bandwagon. You might also check out Craik, Saskatchewan to read of some other sustainable innovations by a local government.

What I am suggesting is that more governance should be carried on locally, with ecocentric issues prominent. In such manner local governments might awaken the Rip van Winkles in bigger political circles. The folks back home must be able to apply more pressures on their political representatives than they are experiencing from the business brontosauruses who are presently their main concern. Local school boards should also have members calling for more attention to the Earth upon which we are all dependent. Governments also depend totally on the Earth; they just don't know it.

Yes, this would constitute a return to Nature, or what is left of it. Our new worldview would make our present one look pretty silly after a while. We would finally have a sensible outlook as an essential component of political and industrial behavior. We would have to surrender the massive superiority complex that estranges us from accepting membership in the Earth community. We might even enjoy living in a world where wealth, dominance, and power would be considered as dumb as they really are. We might help make up governments that refreshingly have a few philosophical convictions to set their sights, and everyone else's, on more admirable goals. At present we are definitely behind the eight ball, hemmed in on the cushion, with a nearly impossible shot needing to be made and very little time left to make it.

We might move forward to a new, saner time in which forests, rivers, lakes, or oceans will become as sacred in the thoughts of people as they once were. Business will be reduced to what is necessary. Corporatism will be recognized as a vicious disease that was once epidemic and passed away when money was no longer worshipped. Life will once more become meaningful.

~~~

[1] Jonathan Swift, *Gulliver's Travels* (England: Wordsworth Editions Ltd., 1992) pp. 184–189.

[2] "We're in Afghanistan to help the Americans, not the Afghans," Murray Dobbin, *The CCPA Monitor,* June 2006, p.34.

[3] *World Watch,* January/February 2007, p.7.

[4] *The CCPA Monitor,* February 2007, p.25.

[5] *The Challenge of Global Warming,* edited by Dean Edwin Abrahamson (Washington, DC: Island Press, 1989), pp.196-209.

[6] *State of the World 2004,* pp.164–179.

[7] *The CCPA Monitor,* November 2006, p. 31, taken from an article in *The Guardian.*

[8] *The CCPA Monitor,* June 2006, p.27.

[9] Arnold Toynbee, *Mankind and Mother Earth* (New York and London, Oxford University Press, 1976) p.588.

[10] "Consumption," *The CCPA Monitor,* March 2007, p.3.
~~~

11: The Fifth Action Principle Spread the Message

"Those who agree with the preceding principles have a duty to spread the word by education and leadership." Most of us realize that before we can bring about change in others we have to change ourselves. Our understanding of the eleven Manifesto principles will depend on our background and on the worldview we held before we read the preceding material. We understand from experience that we must have "readiness" for whatever things we do. We can use a small child as an example, noting that it moves by stages from helplessness to full individuality. We know that there are landmarks in its mobility, crawling, standing, walking, and running, each demanding a new form of agility. In a similar manner we can spread the Manifesto message only to the extent we understand it. In many cases we may want to learn more about some or all of the issues called to our attention.

If the principles of the Manifesto confirm awareness we already have, we are likely to realize that testimony on behalf of Earth integrity is sorely needed. Possessing such realization one can then move to the role of advocacy with understanding that it is important to help bring others to a more conscious role in their own lives. By clarifying our dependence on Earth to people who are completely entrapped in accumulative pursuits, a person can help them to transcend the narrow homocentric view society presently holds. Once individuals begin to think about deeper issues — our impact on the planet being one of them — it will become possible to develop a more ecocentric outlook on life. Conscious awareness that food, air, and water are derived from soil processes will encourage a growing realization that each of us is part of Earth wholeness. We shall perceive more clearly that we are systematically barraged with information intended to dominate our thoughts. As we return to our deepest roots we will realize that they rest in Earth. Much of today's anomie may be cured by return to our instincts — or perhaps by return of our instincts to us. It is a simple and obvious truth that Earth is home place, that it is the taproot of our physical existence. I have read books about people returning to their own roots, which usually means to their birthplace. In a larger sense the Earth substratum everywhere is home, and while I realize the convenience of pavement, I also sense the detachment that it facilitates.

Two Wolves Within

A former student of mine, Joan Hall, a Cherokee by descent, told me this story, which emanated from her families' oral history. It is a story of the message conveyed to a young lad by his Cherokee grandfather.

"There is a terrible fight going on between two wolves that live inside of you," the grandfather told his grandson. "One of the wolves is bad. He is made of envy, sorrow, regret, greed, arrogance, self-pity, guilt, resentment, and false pride. The other one is good. He is joy, peace, love, hope, serenity, humility, kindness, generosity, truth, and compassion."

The boy thought about this for a while and asked, "Which wolf will win the fight, grandfather?"

The old Cherokee replied, "The one you feed."

A problem of that sort resides within us today. The tenor of our times keeps us constantly on the run, with so many things on our agenda that we become fatigued by hyperactivity. There is no time for reflection. We are continually exhorted to buy things, to attend events, to contribute to organizations of various sorts. We have become habituated to interrupting activities when "It's time for the news." In cities, bundles of advertising flyers are left on doorsteps, while a disturbing percentage of newspapers consist mainly of advertisements. Business activities have fostered intrusive omnipresence in our lives, while shopping has become a form of entertainment and sometimes a serious disease. Credit cards have become deadly invitations to suffocating debt. Over the years new holidays have arisen. Many of them become additional occasions when cards should be sent or presents given. This is not belittling sentiment, but is a reminder that the principle of necessity stated by some philosophers is actively fought against by commercialism of all sorts. Some years ago I told those close to me that I wanted no birthday presents, reminding them that I was only born once. I have also reminded people that Christmas should be renamed the Merchant's Holyday.

From the comments above, think of ways in which your pace may become your own and thereby less subjected to outside pressures. Turning off outside media is one of them. Substituting the reading of an informative book instead of watching television is another. Note that when reading, you can pause, lean back, and reflect about the matter that caught your attention. You do not have to pursue a machine to keep up with its story. Devisers of television count on you to suffer through advertisements in order to follow the story. The whole setup is too contrived. A recent university study disclosed that the prime activity of Americans is media consumption, to the tune of nine hours daily. They spend this time on televisions, computers, radios, phones, and music players. Television occupied most of their time, with computers in second place. A Stanford University study also reported that one in eight Americans is addicted to the Internet.[1]

Some people like telephone answering machines because they help to screen callers. We know several individuals who keep their answering machines on constantly to determine whether they want to take an incoming call. More individuals now refuse to participate in "surveys" and reject increasing numbers of telephone solicitations. Technology has made intrusions of privacy more possible. It is harder to forestall commercial invasions of privacy but, as indicated, people are seeking and finding a few ways to keep that nemesis at bay.

Responding to Change

Spreading the Manifesto's principles may require study, personal change, and public activities. I mention study as the first prerequisite because understanding the issues clearly, and understanding how problems have become magnified over time, requires staying informed. There are many books available that address thoughtful issues, such as quieter ways of life, more philosophical outlooks, greater detachment from materialism, and, of course, spiritual matters. It is evident that we are presently discouraged from thinking for ourselves, and that almost every contact we have with media in any form, including personal computers, presents us with an advertising objective. There is an unrelenting effort to keep us focused on trivia or trivial pursuits. Just realizing this is a big step.

How about things we can do? Writing on this topic caused Linda and me to note that not all of our light bulbs are of the energy efficient type. We decided we would correct that matter. We both read regularly so we bought two energy efficient floor lamps that provide a good reading light using 18 watts of energy, and an optional brighter light of 23 watts that enables reading very fine print. Years ago, we developed the habit of turning off lights we were not using. Energy use may be curtailed by improving insulation in homes, by keeping thermostats lower, and quite significantly by turning down hot water heaters. We reduced the temperature of our water heater by twenty degrees from its setting at 140 degrees Fahrenheit when it was installed. We also wash clothing in cold water, using appropriate soap.

So what other things can be done at home? One of them would be to reduce or eliminate electronic entertainment by making a conscious choice of worthwhile reading and study, and making more time for family activity. This leads to a significant reduction of energy consumption. We note that new plasma televisions are great energy wasters. It has been reported that if half of the people in the United Kingdom were to switch over to plasma televisions, it would require the added electrical production of two nuclear generating plants.[2] Linda and I decided years ago that we did not want a television, and as we hear how television shows become cruder and more violent, we are glad that we made that choice. A main objection to television on our part is the intrusion of continual advertising. Since that is TV's purpose we have neither need nor desire for such a waste of time and energy. We do have a cassette and CD player, as we enjoy music.

In the matter of vehicle use, we have been conservative since the mid-1970s. We were somewhat shocked when a neighbor told us that he drives about forty thousand kilometers every year. I think that we were both a bit awed that anyone needed to drive so much. We refrain from lengthy trips. Our vehicle is a 1994 Suzuki Sidekick that we bought in late 1993. Since we bought it we have driven it just over 95,000 kilometers. We go to town once every two or three weeks, a distance of fifty-five kilometers each way. We take few short trips, and when we travel into wilder country several times a year, we often take along a dozen or so seedling trees from our place and plant them in barren spots that look as though they need a tree. Or we might take along a container of seeds and scatter them while we walk: bearberry, huckleberry, clover, or various flower seeds collected from our own flowers at home. We realize that in the long run this act will help to absorb carbon we have introduced to the atmosphere. We also feel that land should be covered with vegetation for its own sake and simply react to that concept. Let's put it this way. We object to the way land is continually laid waste by being altered for the slightest of reasons, and while we know there is a limit to what we can do about it, we choose to do what we can. Gandhi made an accurate comment that whatever we do may not be very important, but it is very important that we do it!

As to other travel, when Linda's mother was in a retirement home in Calgary, Linda would go to visit her regularly by Greyhound bus. The most conservative method of travel is by bus. When it comes to air travel we feel it unnecessary to become tourists. Usually when we are away from home for even a couple days, we are happy to get back. It has a lot to do with one's perspective on life. We note that large numbers of tourists now visit this country. Since we live here already and other people see it as a goal, we feel it is quite appropriate to stay where we are. We keep a pretty accurate record of our walks, and on December 26, 2006 we reached the distance of 600 miles for that year. We enjoy walking in the natural settings around our home. There's nothing particularly wonderful about the distance we cover, but it feels good to be outside and communing with Nature. A day that passes without some sort of walk seems incomplete.

It is apparent that automobile travel is a great danger to planetary health. Ilya Ehrenburg, a Russian journalist commented in 1929 that we couldn't blame the automobile for problems we have caused. Speaking of the auto, he wrote, "It can't be blamed for anything. Its conscience is as clear as Monsieur Citroen's conscience. It only fulfils its destiny; it is destined to wipe out the world."[3] There is no question that many people are addicted to automobiles. The addiction is so complete that, although hints have been given that travel should be rationed, automotive and otherwise, most people are bereft of sufficient willpower to support such a suggestion even though it might mean saving their own lives. A considerable number of years ago, I became a selectman — a very junior sort of politician — in a small town back east. Another selectman, then eighty years old, told me a story from his youth in the days of World War I. It was about an event that had to do with an automobile. At that time there were many people who wanted nothing to do with such new-fangled devices and called them stink buggies. The tale was of one farmer's regular weekend reverie that was interrupted by an automotive incident. I resolved that someday I would try to preserve his story in some way — and this is it:

SOWING AND REAPING

Sunny Sunday mornings found him atop his fence,
Watching autos chugging their ways, hither and thence.
Mostly it baffled him why folks would want such things,
"I s'pose after this the durned fools will want wings."

He chewed on a straw, and stretched with a sigh,
Enjoyed the warmth and watched a flock of geese fly by.
Basking in the peace and quiet of his day of rest,
He heard the sound of footsteps coming from the west.

A broken down stinkbuggy, he thought to himself.
A worthless thing wanted by those with too much pelf.
The quiet was shattered! "How long will it take to town?"
Jaw set grimly, he answered only with a frown.

"Are you deaf? How long will it take to walk to town?"
With the silence of an oyster, he looked down.
The befuddled stranger paraded off down the road,
The fence-sitter eyed him like a recumbent toad.

With distance between them he roared, "Come back here!"
Leaves fell, birds flew, from the woods the bark of a deer.
"Come back!" The walker returned, anger in his face.
The sitter stood and spoke to him with gentle grace.

"I judge that it will take forty minutes walking."
"Why couldn't you tell me without so much balking?"
"You didn't ask me how far to town, but how long;
I had to see how fast you walk to not be wrong."

The stranger took off his cap, and scratched his head,
Shrugged ruefully and felt his face turn red.
"Thanks, I've learned a lot," was all he said,
Then reflectively strode away with measur'd tread.

More Fuels or Less Need for Fuels?

At present, efforts are being made all over the world to plant oil palms for biofuels. These fuels are made from plant or vegetable material. For instance, the city of Stockholm has been using such fuel to provide energy for its maintenance vehicles for some years. Its biofuel is processed from woodchips, sewage, and restaurant wastes and replaces intensive CO_2-producing fossil fuels. Sweden, as many people know, is working steadily toward being completely free of fossil fuels. Biofuels were known about and even used many years ago but only now is the carbon problem making them an attractive alternative. Sadly, vast acreages of natural, diverse forests have been destroyed by burning to allow mass planting of oil palms, which are rapidly growing trees suitable for biofuels. With typical obtuseness a new economic opportunity is envisaged and is now being pursued in this way on millions of hectares in sub-tropical or tropical latitudes. Indonesia and Malaysia are two such places where natural forests and the biodiversity they supported were burned on money's altar just to sustain human mobility. Numerous schemes for powering machines are afoot, but almost no effort or thought is being put into redirecting our ideas toward great reduction of vehicle use or of the need for reducing transportation in general.

By applying a value judgment to intended mass production of biofuels we can say that, within limits, they are preferable to fossil fuels because biofuels produce relatively miniscule amounts of CO_2. In a homocentric sense, entrepreneurs see them as profitable. But mass-production of biofuels at the expense of natural biodiversity is blasphemy. The central issue that is being ignored is that the amount of travel, worldwide, should be vastly reduced. According to George Monbiot we must reduce our carbon emissions by 90 percent.[4]

Not too many years ago, Ralph Waldo Emerson, in his essay "Self-Reliance" categorized travel as a fool's paradise. I recall this particularly because of an alumni publication I receive that offers in every issue an endless recounting of places visited by fellow alumni. When I was in secondary school I chose Emerson's essay "Self-Reliance" as a required report on a non-fiction topic. Perhaps I was influenced at an early age by Emerson to throw out my anchor in some quiet, peaceful place and stay there. Here is a short excerpt:

> Traveling is a fool's paradise. Our first journeys discover to us the indifference of places. At home I dream that at Naples, at Rome, I can be intoxicated with beauty and lose my sadness. I pack my trunk, embrace my friends, embark on the sea and at last wake up in Naples, and there beside me is the stern fact, the sad self, unrelenting, identical, that I fled from. I seek the Vatican and the palaces. I affect to be intoxicated with sights and suggestions, but I am not intoxicated. My giant goes with me wherever I go.
>
> But the rage of traveling is a symptom of a deeper unsoundness affecting the whole intellectual action. The intellect is vagabond, and our system of education fosters restlessness. Our minds travel when our bodies are forced to stay at home. We imitate; and what is imitation but the traveling of the mind?[5]

A number of years ago I was visited by a couple from Chicago who had recently made a tour of art museums in several European countries. They had many pictures, which they showed me, some of which were interesting. After looking at their pictures, Joe asked me whether or not I would enjoy such a tour. I remarked that there was quite endless art visible around where I live, and that I had traveled sufficiently when I was overseas in the Second World War. Several months later I received a letter from them with a cartoon enclosed that showed two mounted cowboys in a spectacular canyon area. They sent it along, they wrote, because it reminded them of me. One cowboy was saying to the other, "Yup, I'd rather live in a masterpiece than own one." I probably have that cartoon someplace. Anyway it did fit my sentiments quite accurately.

Is my insertion of the Emerson observation and the cowboy cartoon incident an attempt to stir your own cogitation? Sure. Why not? There are reams of travel articles in all sorts of publications. Frankly I think they are a waste of trees, but if you are interested in any particular area of the world you might learn more about it by reading and studying than by taking a packaged tour. Many travel videos or DVDs are also available for rent. People everywhere need to examine their own restlessness, their own desire to imitate others, and at bedrock ask themselves how much of our incessant chasing from place to place can be afforded by our planet. Violent weather should be telling us to cut back on fossil fuel use. Are, or are we not willing to change our ways for the sake of our children and grandchildren whose fate depends on the way we are stacking the cards? I may not be as effusive about loving humanity as some are, but it seems that claiming to love others without willingness to make necessary changes to preserve their lives is pretty thin whitewash that won't last very long.

The carbon problem and solutions exemplify an emergency message that needs to be spread. Developed countries have been on a long-sustained drunk. This analogy is appropriate to increasing automobile and aircraft use. Such a drunken spree must come to an end. Who says so? Since the planet is part of an orderly cosmos, we can easily reflect that the universe says so.

Carbon Was Stored for a Purpose

Basically, the problem is simple. Earth evolved its own ways of storing carbon. Man extracted stored hydrocarbons for what has turned out to be insatiable use. Then he learned that there were consequences. But he wanted to go faster and faster, farther and farther, more and more. He became a plague on the planet. Nature has been steadily issuing warnings.

Man wants to ignore them and to go even faster by railway trains, by faster aircraft, by playing with rocketry, by shipping sports teams regularly from place to place, by creating galas, Olympics, festivals, and other occasions to entertain ourselves. We push against Nature's warnings, strain against the fence; refuse to acknowledge the abyss it hides.

We may thus be considered a rigidly recalcitrant species. Many of us will use any form of denial we can think of to justify our continued random burning of fossil fuels. These people are so addicted that they will desperately depend on the slogan "They will find a way." There is a way! It is to ration fossil fuel use drastically. Instead of seeking to find more fuels, learn to be modest consumers of fuels. Stay home! Cure our addiction to machines. Learn to walk again, and to enjoy walking. If you want to spread a meaningful message to others, it could be to keep their feet on the ground. Flying anywhere is a menace to Earth. Aircraft, particularly jet aircraft, should be avoided. The funny thing is that we don't really need any of them. When we examine the damage aircraft travel is doing to our planet, it is doubtful whether any person on Earth is important enough to have to fly anywhere. Our sense of importance might be diminished, but reality would be a bit closer.

Yes, there are all sorts of things we can do to spread the message of caring for the planet, of respecting it as the cause and sustenance of our being, and of seeking to live in harmony with it. We need to make a testimony of our own lives and try to live more peacefully and thoughtfully. Many have succumbed to a foolish doctrine that would have us run full speed from one activity to another, and live in servitude to machines. Society has enslaved itself willingly to these machines and has surrendered its collective soul to haste, waste, and a ridiculous decision to make production and material wealth its goal. Lives of ceaseless activity constitute a denial of the spiritual core of being that has been identified as our essence by the labors and thoughts of countless philosophers, historians and theologians. The problem is not that we made the machine but that we let the machine remold us in its own image.

If you have read the book up to this point, you are probably aware that I am not optimistic about the future of humanity. I am also about as positive as one can be that world leadership is hopelessly committed to big money, big industry, and the pursuit of profit. It seems paralyzed by addictive double ignorance. Ted Mosquin's considered opinion is "that we are witnesses to an emerging ecological nightmare in which pessimists are realists and the optimists are deluded. The hard evidence from trends the world over cannot be interpreted in any other sensible way." Sometimes I ask myself just where I stand in regard to the fate of our species. As I ponder this issue I realize that civilization is an ideal in my own mind. I also realize that it is not the trappings of civilization that I admire, but the casual friendliness of people not deluded by illusions of their importance, people who do not cultivate pedigrees, people whose own spiritual drives are toward greater understanding of life and its deeper meanings. Such aspirations do not necessarily make a socialite of a person but may move an individual toward simple pursuits and indeed toward reverence for life as a sufficient cosmology. I would like to see our species save itself by subordinating its expectations to what the Earth can easily provide.

Stoics and the Cardinal Virtues

I have not been able to find a more pertinent outlook on life than was presented in the four divisions of virtue established by the Stoics: temperance, fortitude (or courage), prudence, and justice. These virtues might lead to a sort of "don't expect much — don't demand much" way of life. In this manner of living, sunlight is more important than limelight. I have said before and say now, that following the four virtues is about the only alternative I can envisage that might save our species. Religions tend to shake their fingers at the "seven deadly sins," and get a lot of oratorical mileage lecturing against pride, avarice, greed, envy, lust, anger, gluttony, and sloth. But avarice, which we commonly call greed, has become the motive that drives the economy, and pride has recruited many a susceptible buffoon.

But how often do churches focus on the positive effect of cultivating virtues that are capable of neutralizing and overpowering sin? How often do churches approach William Blake's ability to look behind the veil of appearances, "To see a world in a grain of sand/ And a heaven in a wild flower, Hold infinity in the palm of your hand/ and eternity in an hour"? How many people can realize that "A robin redbreast in a cage/ Puts all Heaven in a rage" or that "A dog starv'd at his master's gate/ Predicts the ruin of the state."

If anything can return us to the path toward civilization it will be cultivation of principles derived from the virtues that society has apparently deemed meaningless and old-fashioned.

I suggested in an earlier book, *The Soul Solution,* that we stand at a time in history that resembles the pre-Stoic era in which people had lost faith in the Roman and Greek gods (about 300 BC). Educated Greeks asked philosophers for a new worldview that would comfort people and give meaning and value to life, and also reduce the anguish people often feel at the thought of death. The philosophy of Stoicism came into being and developed a dogma that is much needed today. This philosophy called for ascetic lives based on simplicity, morality, and a belief that all things are within Divinity. Actually Stoicism was an important root of Christianity. Christianity should review its own Stoic roots for the effect Stoic thought might have in regard to protection of the Earth. Materialistic, capitalistic Christianity, as we know it today, seems anything but spiritual.

The part of Stoicism that fits biblically with the Genesis concept that we were put on Earth to dress and keep it, is superior to the virtual disdain that the majority of people exhibit toward the planet and its integrity.

A Theology of the Earth

As I did in *The Soul Solution,* I once again suggest the need for a theology of the Earth, both within existing religions and outside them. Believers and non-believers are both needed to overcome the Faustian pact with the "money devil." Instead of lamenting, striving, fighting, and killing for greater wealth and possessions, thus making life a form of daily warfare, we need to modify our expectations and be grateful for modest possessions. We must begin to appreciate those intangible qualities of life that more than compensate for the conspicuous waste that has replaced fresh, clean air, abundant water, and unpolluted soil. We have traded good health for suppressed immune systems and man-made afflictions. There really is pleasant wealth in not wanting things.

The Stoics were on the right track when they spoke of the universe as a living organism with Divinity as the soul, the source of universal law, and the infusing spirit that sustains all. Thoughtful life in natural surroundings still imparts a strong sense of serenity. Morality, the Stoics further thought, is willing surrender to the laws of the universe. Religious doctrines have expanded on these ideas and instituted themselves as necessary vectors for attaining their specified after-life. Their continuance depends on being accepted as such vectors. The fact that there are thousands of religions claiming to be the "only way" to believe, still reminds me that life will be best sustained by adherence to the ecological organization of this planet. Such adherence is essential. I find it sensible to think that Heaven is under our feet as well as overhead.

Stoicism indicated virtue as the highest good, a contention espoused by many thinkers. Today we spurn such an idea as simplistic, and support cutthroat competition as the praiseworthy path to follow. I would like to illustrate how the implementation of these virtues would enable us to move rapidly toward a more hopeful and more sustainable world than we now have.

The Four Cardinal Virtues

I will start with temperance, because I read somewhere in the writings of Michel de Montaigne that when he was thirteen years old he was required to read and discuss Aristotle's essay on temperance. That led me to look up what Aristotle had to say on the topic and I found the essay easy enough to understand. I feel that young people who are good readers, with the help of discussion, could digest Aristotle's meaning. Armed with this virtue, individuals would be able to neutralize the impact of consumer advertising. If enough people became temperate, our politicians, business leaders, and media forces would at first be bewildered, next they would be educated, and thirdly, they would themselves, necessarily become philosophical. "I don't want it; I don't need it; I have enough." The adoption of the virtue of temperance would bring conspicuous consumption to a shattering stop. Immediately someone will ask, "What about jobs?" Yes, there would be an impact on jobs; but many more jobs would be created than would be lost when we begin to clean up the planet from the mess we have made of it. We would bring our armies home to help necessary work and arrange for politicians, if they must fight, to fight with each other since they are historically the instigators of wars.

Temperance is simply the modification of expectations, the desire to live a modest, unassuming life. It could become an excellent topic for discussion, an excellent subject for teaching, an excellent way of living. Temperance would restore a human pace to life. As an attitude it would bring up the question, "What are we all hurrying about?" I realize that these are not temperate times. Industrial forces pulled out our throttles and we are now trying to live on the dead run. However it is not that difficult to change.

Yes, temperance pertains to all sorts of things: to how much one eats, drinks, works, loafs, reads, walks, runs, sleeps, and just about any activity that a person undertakes. Temperance infers moderation in all things. And, believe it or not, one must occasionally be temperate about temperance. Temperance then is best known as the ability to satisfy limited desires, whereas intemperance is the plague of boundless desire that can never be sated. It does not require intense study of modern society to recognize a perpetual encouragement toward intemperance, advertising being its chief advocate.

The second of the cardinal virtues we will consider is fortitude, which we now usually call courage. This virtue depends as much on strength of mind as it does on strength of body. The sort of courage now being called upon in these times is the kind necessitated by sacrifice of valued activities and pleasures to protect the well-being of all life forms on Earth and the integrity of the planet itself. St. Thomas Aquinas (1225–1274) pointed out the interrelationship of the virtues by explaining that courage overflows into the rest of the virtues, as they in turn enter into courage. Aristotle contended that it is "for a noble end that the brave person endures and acts as courage directs." While unthinking persons might say they are not going to give up any of their pleasures — driving, flying or whatever the pleasures might be — they ignore the generations of young people who are continually being born into a world endangered by human actions. Refusing to consider our obligation to other generations is not only a form of denial but a lack of sufficient courage to squarely face the dark abysses of the times in which we live. The times we live in call for change. We know that deep inside ourselves. Do we have the courage to sacrifice our desires for the good of all? At this point is should be clear why temperance is also a mode of courage.

Prudence must also be employed. In origin, there are two related Latin words. *Scientia* (science) means knowledge and *prudentia,* or prudence, is more than knowledge. For example, through science we have devised the means of making nuclear bombs. Prudence involves not only intellectual qualities, but also moral qualities. As well as knowledge, it calls into play the quality of experience, reason, and moral judgment, which would enable us to see clearly how worldwide nuclear wars might result from making nuclear bombs. Science, used without prudence, has widely poisoned our entire environment. Toxic metals now appear in the flesh of most animals including fish, birds, reptiles, amphibians and mammals. The prudence that we need to prevent disaster is lacking. Prudence might interfere with our shortened definition of progress, so our madcap economy simply discards such a hindrance. Some analysts describe prudence and wisdom as synonymous. Others believe that all four cardinal virtues must be cultivated and that true wisdom is received only as a gift of the gods. Either way, modern problems reflect a lack of prudence/wisdom.

The fourth virtue, justice, has often been described as the "interest of the stronger," or the view that "might makes right." Governing bodies often adopt such a tempting form of justice. As a result, society is often permeated with dictates of the mighty. This form of justice allows government to be above the law; but a higher form of justice concludes that governments themselves should be held accountable for their actions. Superior to both these definitions of justice, there is a third form called "original justice." Theologian Thomas Aquinas interpreted this highest form as stemming from one's "reason being subject to God," and "the lower powers subject to reason." I would like to note here the obvious relationship between ecocentric thought and original justice. Ecocentrism suggests living in harmony with natural law, which therefore involves justness of behavior that attempts to harmonize itself to the "great wholeness" in which we have our Being. This amounts to original justice as closely as we can understand it. Once reason extends itself beyond narrow homocentric limitations, it is evident that we have a long history of perpetuating injustice towards the animals and plants that are co-inhabitants of the ecosphere. Bedrock morality demands of us that original justice must be soberly extended to Earth and its entire panoply of inorganic structures and living organisms.

The cardinal virtues are given here because it seems sensible that we must change our thinking before we can change our ways. Hoping that we can modify our way of thinking in time, adherence to these virtues would help suppress the evils of pride and greed, which have forced us to the brink of extinction.

It is not false to say that we are there — at the edge of disaster or renewal — and it will be the latter if we don't make it the former.

[1] *ecologist,* January 2007, p.038.

[2] *ecologist,* January 2007, p.009.

[3] *Heat,* p.142

[4] *Heat,* p. 173.

[5] Ralph Waldo Emerson, *The Selected Writings of Ralph Waldo Emerson* (New York: Random House, 1950) p.165.

Afterword

Writing this book has been an attempt to reach into the thoughts of readers and convince them that we must give the needs of our planet priority over our own wants. We are all earthlings, right down to the bottom-line profiteers and most powerful politicians. We have called our book *Testimony for Earth* because Earth deserves our testimony more than anything else we know. Unfortunately there is downright little concern, and even less respect, for the planet that underwrites our birth and sustenance throughout our lives.

Through constant exhortations stemming from world leaders of all sorts, business leaders, politicians, religionists, and others holding special interests, we have been vigorously conditioned to believe that nothing but humanity is of any importance. Our continuance as a viable species depends on rejecting that illusion and making an abrupt shift from self-serving homocentric concerns to planet-saving ecocentric ones. This boils down to a change in focus. The ideas presented in preceding pages call our homocentrism into question by presenting evidence that our conviction of absolute independence and importance is an illusion.

A March 21, 2007 Canadian Press news story written by Chinta Puxley is headlined, "Rich, educated Cdns won't Give up SUVs: poll." The article reflected the findings of an on-line Angus Reid poll that disclosed varied Canadian attitudes toward climate change. The most significant one may be that one-third of Canadians "consider climate change to be the most important issue facing humanity." Chief executive Angus Reid noted that the only time Canadians showed such environmental concern was in the late 1980s when fears about acid rain and overflowing landfills led to the blue box recycling programs with which we have become familiar. Not only has environment moved back into first place among our concerns, it also has huge staying power said Reid.

The survey shows that well educated and wealthy Canadians are the most disinclined to change their ways in order to improve the environment. A majority of people who responded to the poll were reluctant to drive more fuel-efficient vehicles, to lower thermostats in their homes, to reduce their consumption of hot water, or to reduce their amount of air travel. These respondents minimized the impact of Canadians on climate and pointed their fingers at China and India as bigger polluters. Threats to affluent lifestyles are basically rejected.

The respondents most concerned about the environment were from Quebec. Three-quarters of respondents from Ontario were convinced global warming is taking place, and half worried that their lives would be significantly impacted.

In spite of peoples' concern, Chris Winter head of the Conservation Council of Canada, commented that there is evidence all around that people aren't taking things seriously. Hummers and SUVs can be seen, and people have lights on at all hours of night, he observed.

It seems that the poorer segment of society is much more concerned than the more well-to-do, and that many people are hoping that governments will do something to force very significant change.

The Earth has been our kindly sponsor. It has tolerated assaults on its integrity to such an extent that we have made the dangerous assumption that we can do anything we want and get away with it. That is patently false. A host of Earth warnings ranging from severe earthquakes to rambunctious weather events are taking place successively as Earth clearly informs us that we have overstepped our capabilities in numerous ways. There is much evidence that corroborates a series of truths: the Earth is our wealth; the rabid economy, once thought to be a glorification of human intelligence has impoverished us by its insatiability; and, willing, rigid obedience to Earth's ecological requirements is our sole hope for survival.

Truthfully, we are in a situation that should evoke an all out effort. In spite of the array of power and might we have concocted, we are fragile lumps of protoplasm, childishly asserting our divine rights in the face of a colossus that could wipe us all out in a few moments.

Elsewhere in the world, things bad and good are happening. NASA's top scientist is quoted in an editorial in the March/April issue of *World Watch* magazine as saying that we have "perhaps 10 years to get a grip on our carbon emissions or else face a tipping point beyond which climate warming spirals irreversibly out of control."[1] Other scientists have indicated doubt that we will have that long. A serious problem is that, given the possibility of ten years before catastrophe, politicians will equivocate for eight or nine years before trying to do anything that is really sensible.

The same editorial called urgent attention to the rush in Texas to build 150 "new" pulverized-coal power plants before likely federal statutory limits on carbon emissions make such plants illegal. This eighty-year-old technology "is hardly more advanced than a log fire." If the plants are built, they will release "billions of tons" of non-sequestered carbon into the air in their fifty-year lifespan.

Editorialist Thomas Pugh suggests in biting terms that such coal-fired plants fit the role of "carbon crimes," which need to be outlawed. He fantasizes that the rush to build such facilities before law forbids them should lead to serious criminal penalties. "The guilty utility executives would live in un-air-conditioned, unlit cells and spend their waking hours planting trees in the tropics…sweating under the hot sun to undo some small part of the harm they'd caused." I think this is a significant commentary on the fact that people are looking with extreme contempt at fanatical development that will lead to the chaotic collapse of society.

Things are also happening on the good side in the race between endless greed and more sensible behavior.

"Cuban cities grew more than one million tons of vegetables and spices between January and March of 2006." The food was produced on more than 35,000 hectares of container gardens, small plots, and some high-production gardens that produced a variety of crops. This was primarily organic gardening. Three hundred fifty thousand jobs were created, of which 20 percent were filled by women. Much as Cuba has been criticized, notably by the US, the Castro government has been literally more down to earth. Because of the small size of the garden plots, they suffered minimal damage from several hurricanes that occurred. Fossil fuel use was also reduced because food did not have to be transported long distances to reach consumers.[2] Although Canada does not have Cuba's favorable climate, I am sure that with modification of methods, much food could be produced in Canadian cities. This would reduce the 2,000 or more miles that many foods must be transported to Canadian tables, and would produce food that is free from pesticides. We could commence getting down to earth in minds and hearts, and develop the initiative to use all bits of available land for food production. We would also regain at least some wee fragments of our instinct and energy by contact with the Earth, and learn to provide, to some extent, for ourselves. Beginnings such endeavors may be difficult but may in time lead to greater things.

An article by Helke Ferrie recommends a major shift from agribusiness to organic farming. Reason enough is that "20 percent of all greenhouse gases in the world arise" from modern industrial agriculture. The author, a medical science writer, mentions a Cree Indian prophecy that seems to predict the future clearly: "Only after the last tree has been cut down. Only after the last river has been poisoned. Only after the last fish has been caught. Only then will you find that money cannot be eaten."[3]

Some people will remember a song with a lyric that suggested, "Enjoy yourself, enjoy yourself. It's later than you think." Society has been acting that way for decades. Travel, seeking pleasure, and accumulating possessions have become our way of life. We have also been aware that we are over-stressing the health of earth, but have constantly denied that truth. The same denial is carried on today because we lack the will to face our problems. "There will be technological solutions," we blindly contend, and on that fragile hope we continue abusing the ecosphere. We rely on science without realizing that the planet's precarious health is a consequence of uncontrolled technological disintegration of ecosystems. While science itself may be described as neutral, ecosystem-destroying, industrial science is anything but neutral and is cynically employed for profit.

Paradoxically, the Canadian government recently announced that it would contribute $30 million toward the Great Bear Rainforest in British Columbia and thereby protect several species of bears. I was puzzled to hear that it would cost $30 million to leave something alone. I couldn't afford that on my out-of-the-way eighty acres, which cost $7,000 thirty-five years ago. It has remained a fair representative of wilderness by mainly leaving it to Nature's care. She is the best custodian one can have.

We are standing at an axial point in history. There are alternatives. One is that we can continue along our merry way, and let ourselves be manipulated, and ultimately thereby condemn our children, grandchildren, and our entire species. The other is that we can think of Canada in an ecocentric way. This means that we would "stand very thoughtfully on guard" for Canada. That is to say, for the health of the land.

I definitely do not believe that politicians have the fortitude to save us from the fate we are inviting. They are too used to compromise, too subservient to industry, and too concerned with re-election to trample on the toes of the industrial fanatics who have run the world to suit themselves for an almost fatally long while. The common people, I believe, necessarily remain a bit closer to Earth than the movers and shakers who have matriculated into the power structure. By unleashing the power inherent in the masses, the surrendered values of society could be restored. Just by *not* doing things we can make a big difference.

For instance, we can help the planet by reducing or eliminating our consumption of meat. A recent report from the UN came to some stunning conclusions about meat consumption. "The livestock sector emerges as one of the top two or three most significant contributors to the most serious environmental problems, at every scale from local to global...The UN report says almost a fifth of global warming emissions come from livestock."[4] That is a greater amount of global heating gases than is produced by transportation. Much evidence also exists that our society consumes amounts of meat detrimental to its health.

We need to stop looking only inward at things we want for ourselves, and look outward at things we should really stand on guard for — such as the beauty and natural order in which we have been given the gift of life — possibly many times.

Our option is to be content with here and now, and to realize it is much more peaceful to learn to live adequately near home rather than to chase all over the world looking for yourself. I am reminded of an individual I knew who was lost in Calcutta. He stopped a man who, fortunately, could speak English. He held out a map he had purchased and asked the man, "Where am I?" The man ignored the map, looked at him penetratingly, studied him for a moment, then stamped his foot on the ground and said, "Right here!"

In the final analysis our human fate will be the consequence of millions of individual decisions. Many of the decisions may appear to be minor issues, but these can add up to monumental change.

Let's hope that Earth, with or without us, abides forever!

[1] "Carbon Crimes," *World Watch,* Vol. 20, No. 2, March/April (2007), p.2.
[2] Ibid, "Urban Agriculture Provides Cubans with Food, Jobs," p.7.
[3] Ferrie, Helke. "Key to saving our planet: convert agribusiness to organic farming," *The CCPA Monitor,* September 2005, p.12.
[4] "Vegetarian the way to go," *The CCPA Monitor,* March 2007, p.26.

A Manifesto for Earth

By

Ted Mosquin

and

J. Stan Rowe (June 11, 1918 to April 6, 2004)

This Manifesto has been published in the quarterly journal: *Biodiversity* Volume 5, No. 1, pages 3–9, January/March 2004. The journal is owned by The Tropical Conservancy, a charitable organization whose address is 94 Four Seasons Drive, Ottawa, Ontario, K2E 7S1. Subscriptions rates and back issues at URL: <http://www.tc-bioidiversity.org/> where an electronic version of the Manifesto will be available in the near future. A pdf file of the Manifesto (with graphics) can be downloaded at www.ecospherics.net/ pages/EarthManifesto.pdf

About the Authors

Ted Mosquin has a Ph.D. in Systematics & Evolution from UCLA. He spent 12 years as a research scientist with Agriculture Canada, Ottawa, and has taught at the University of Alberta, and the University of California, Berkeley. He served as editor of The Canadian Field - Naturalist, and of Biodiversity. He is the author or co-author of four books and of some 100 scientific and popular articles on ecology, natural history, endangered species, biodiversity, and environmental ethics. A recent article: The Roles of Biodiversity in Creating and Maintaining the

Ecosphere http://www.ecospherics.net/pages/MosqEcoFun5.html summarizes part of the ecological foundation for this Manifesto. Ted has served as President and Director of several national and regional Canadian environmental organizations (More bio details at: www.ecospherics.net/pages/aboutauthors.html).

Stan Rowe was educated in ecology at the Universities of Alberta, Nebraska and Manitoba. He has spent equal time as a research forester with Forestry Canada, as a teacher at the University of Saskatchewan, and since 1985 as an emeritus Professor. A geo-ecologist and environmental ethicist with a background in silviculture and terrain (landscape) ecology, Stan authored *Forest Regions of Canada* (1959), *Earth Alive: Essays on Ecology* (NeWest Press, Edmonton, 2006), and *Home Place: Essays on Ecology* (NeWest Press, Edmonton, 1990; reissued 2002), as well as numerous articles, book chapters and reviews. Some of his articles on ecology and ethics are posted at www.ecospherics.net. He has served on provincial and federal environmental advisory councils. (More bio details at: http://www.ecospherics.net/ pages/aboutauthors.html).

Preamble

Many artistic and philosophical movements have produced Manifestos, proclaiming truths that to their authors were as manifest as their five-fingered hands. This Manifesto also states self-evident truths, as obvious to us as the marvellous five-part environment - land, air, water, fire/sunlight, and organisms - wherein we live, move, and have our being. The Manifesto is Earth-centered. It shifts the value-focus from humanity to the enveloping Ecosphere - that web of organic/inorganic/symbiotic structures and processes that constitute Planet Earth.

The Ecosphere is the Life-giving matrix that envelops all organisms, intimately intertwined with them in the story of evolution from the beginning of time. Organisms are fashioned from air, water, and sediments, which in turn bear organic imprints. The composition of sea water is maintained by organisms that also stabilize the improbable atmosphere. Plants and animals formed the limestone in mountains whose sediments make our bones. The false divisions we have made between living and non-living, biotic and abiotic, organic and inorganic, have put the stability and evolutionary potential of the Ecosphere at risk.

Humanity's 10,000-year-old experiment in mode-of-living at the expense of Nature, culminating in economic globalization, is failing. A primary reason is that we have placed the importance of our species above all else. We have wrongly considered Earth, its ecosystems, and their myriad organic/inorganic parts as mere provisioners, valued only when they serve our needs and wants. A courageous change in attitudes and activities is urgent. Diagnoses and prescriptions for healing the human-Earth relationship are legion, and here we emphasize the visionary one that seems essential to the success of all others. A new worldview anchored in the planetary Ecosphere points the way.

Statement of Conviction

Everyone searches for meaning in life, for supportive convictions that take various forms. Many look to faiths that ignore or discount the importance of this world, not realizing in any profound sense that we are born from Earth and sustained by it throughout our lives. In today's dominating industrial culture, Earth-as-home is not a self-evident percept. Few pause daily to consider with a sense of wonder the enveloping matrix from which we came and to which, at the end, we all return. Because we are issue of the Earth, the harmonies of its lands, seas, skies and its countless beautiful organisms carry rich meanings barely understood.

We are convinced that until the Ecosphere is recognized as the indispensable common ground of all human activities, people will continue to set their immediate interests first. Without an ecocentric perspective that anchors values and purposes in a greater reality than our own species, the resolution of political, economic, and religious conflicts will be impossible. Until the narrow focus on human communities is broadened to include Earth's ecosystems - the local and regional places wherein we dwell - programs for healthy sustainable ways of living will fail.

A trusting attachment to the Ecosphere, an aesthetic empathy with surrounding Nature, a feeling of awe for the miracle of the Living Earth and its mysterious harmonies, is humanity's largely unrecognized heritage. Affectionately realized again, our connections with the natural world will begin to fill the gap in lives lived in the industrialized world. Important ecological purposes that civilization and urbanization have obscured will re-emerge. The goal is restoration of Earth's diversity and beauty, with our prodigal species once again a cooperative, responsible, ethical member.

~~~

Core Principles

Principle 1. The Ecosphere is the Center of Value for Humanity
Principle 2. The Creativity and Productivity of Earth's Ecosystems Depend on their Integrity
Principle 3. The Earth-centered Worldview is supported by Natural History
Principle 4. Ecocentric Ethics are Grounded in Awareness of our Place in Nature
Principle 5. An Ecocentric Worldview Values Diversity of Ecosystems and Cultures
Principle 6. Ecocentric Ethics Support Social Justice

Action Principles

Principle 7. Defend and Preserve Earth's Creative Potential
Principle 8. Reduce Human Population Size
Principle 9. Reduce Human Consumption of Earth Parts
Principle 10. Promote Ecocentric Governance
Principle 11. Spread the Message

~~~

Why this Manifesto?

This Manifesto is Earth-centered. It is precisely ecocentric, meaning home-centered, rather than biocentric, meaning organism-centered. Its aim is to extend and deepen people's understanding of the primary life-giving and life-sustaining values of Planet Earth, the Ecosphere. The Manifesto consists of six Core Principles that state the rationale, plus five derivative Action Principles outlining humanity's duties to Earth and to the geographic ecosystems Earth comprises. It is offered as a guide to ethical thinking, conduct and social policy.

Over the last century advances have been made in scientific, philosophical and religious attitudes to non-human Nature. We commend the efforts of those whose sensitivity to a deteriorating Earth has turned their vision outward, to recognition of the values of the lands, the oceans, animals, plants and other creatures. And yet, for lack of a common ecocentric philosophy, much of this goodwill has been scattered in a hundred different directions. It has been neutralized and rendered ineffective by the one, deep, taken-for-granted cultural belief that assigns first value to Homo sapiens sapiens and then, sequentially, to other organisms according to their relatedness to the primary one.

The recent insight that Earth, the Ecosphere, is an object of supreme value has emerged from cosmologic studies, the Gaia hypothesis, pictures of Earth from space, and especially ecological understanding. The central ecological reality for organisms - 25 million or so species - is that all are Earthlings. None would exist without Planet Earth. The mystery and miracle called life is inseparable from Earth's evolutionary history, its composition and processes. Therefore, ethical priority moves beyond humanity to its inclusive Earth home. The Manifesto maps what we believe is an essential step toward a sustainable Earth-human relationship.

Core Principles

Principle 1: The Ecosphere is the Center of Value for Humanity

The Ecosphere, the Earth globe, is the generative source of evolutionary creativity. From the planet's inorganic/organic ecosystems organisms emerged: first bacterial cells and eventually those complex confederations of cells that are human beings. Hence, dynamic ecosystems, intricately expressed in all parts of the Ecosphere, exceed in value and importance the species they contain.

The reality and value of each person's ecological or outer being has attracted scant attention compared to the philosophic thought lavished on humanity's inner being, the latter an individualistic focus that draws attention away from ecological needs and neglects the vital importance of the Ecosphere. Extended to society as concern only for the welfare of people, this homocentrism (anthropocentrism) is a doctrine of species-selfishness destructive of the natural world. Biocentrism that extends sympathy and understanding beyond the human race to other organisms marks an ethical advance, but its scope is limited. It fails to appreciate the importance of the total ecological "surround." Without attention to the priority of Earth-as-context, biocentrism easily reverts to a chauvinistic homocentrism, for who among all animals is commonly assumed to be the wisest and best? Ecocentrism, emphasizing the Ecosphere as the primary Life-Giving system rather than merely life's support, provides the standard to which humanity must appeal for future guidance.

We humans are conscious expressions of the Ecosphere's generative forces, our individual "aliveness" experienced as inseparable from sun-warmed air, water, land, and the food that other organisms provide. Like all other vital beings born from Earth, we have been "tuned" through long evolution to its resonances, its rhythmic cycles, its seasons. Language, thought, intuitions - all are drawn directly or metaphorically from the fact of our physical being on Earth. Beyond conscious experience, every person embodies an intelligence, an innate wisdom of the body that, without conscious thought, suits it to participate as a symbiotic part of terrestrial ecosystems. Comprehension of the ecological reality that people are Earthlings, shifts the center of values away from the homocentric to the ecocentric, from Homo sapiens to Planet Earth.

Principle 2: The Creativity and Productivity of Earth's Ecosystems Depends on their Integrity.

"Integrity" refers to wholeness, to completeness, to the ability to function fully. The standard is Nature's sun-energized ecosystems in their undamaged state; for example, a productive tract of the continental sea-shelf or a temperate rain forest in pre-settlement days when humans were primarily foragers. Although such times are beyond recall, their ecosystems (as much as we can know them) still provide the only known blueprints for sustainability in agriculture, forestry, and fisheries. Current failings in all three of these industrialized enterprises show the effects of deteriorating integrity; namely, loss of productivity and aesthetic appeal in parallel with the continuing disruption of vital ecosystem functions.

The evolutionary creativity and continued productivity of Earth and its regional ecosystems require the continuance of their key structures and ecological processes. This internal integrity depends on the preservation of communities with their countless forms of evolved cooperation and interdependence. Integrity depends on intricate food chains and energy flows, on uneroded soils and the cycling of essential materials such as nitrogen, potassium, phosphorus. Further, the natural compositions of air, sediments, and water have been integral to Nature's healthy processes and functions. Pollution of these three, along with exploitive extraction of inorganic and organic constituents, weakens ecosystem integrity and the norms of the Ecosphere, the fount of evolving Life.

Principle 3: The Earth-centered Worldview is Supported by Natural History

Natural History is the story of Earth unfolding. Cosmologists and geologists tell of Earth's beginnings more than four billion years ago, the appearance of small sea creatures in early sediments, the emergence of terrestrial animals from the sea, the Age of Dinosaurs, the evolution with mutual influences of insects, flowering plants, and mammals from which, in recent geological time, came the Primates and humankind. We share genetic material and a common ancestry with all the other creatures that participate in Earth's ecosystems. Such compelling narratives place humanity in context. Stories of Earth's unfolding over the eons trace our coevolution with myriad companion organisms through compliance, and not solely through competitiveness. The facts of organic coexistence reveal the important roles of mutualism, cooperation, and symbiosis within Earth's grand symphony.

Cultural myths and stories that shape our attitudes and values tell where we came from, who we are, and where in the future we are going. These stories have been unrealistically homocentric and/or other-worldly. In contrast, the evidence-based, outward-looking narrative of humanity's natural history - made from stardust, gifted with vitality and sustained by the Ecosphere's natural processes - is not only believable but also more marvelous than traditional human-centered myths. By showing humanity-in-context, as one organic component of the planetary globe, ecocentric narratives also reveal a functional purpose and an ethical goal; namely, the human part serving the greater Earth whole.

Principle 4: Ecocentric Ethics are Grounded in Awareness of our Place in Nature

Ethics concerns those unselfish attitudes and actions that flow from deep values; that is, from the sense of what is fundamentally important. A profound appreciation of Earth prompts ethical behavior toward it.

Veneration of Earth comes easily with out-of-doors childhood experiences and in adulthood is fostered by living-in-place so that landforms and waterforms, plants and animals, become familiar as neighborly acquaintances. The ecological worldview and ethic that finds prime values in the Ecosphere draws its strength from exposure to the natural and semi-natural world, the rural rather than the urban milieu. Consciousness of one's status in this world prompts wonder, awe, and a resolve to restore, conserve, and protect the Ecosphere's ancient beauties and natural ways that for eons have stood the test of time.

Planet Earth and its varied ecosystems with their matrix elements - air, land, water, and organic things - surrounds and nourishes each person and each community, cyclically giving life and taking back the gift. An awareness of self as an ecological being, fed by water and other organisms, and as a deep-air animal living at the productive, sun-warmed interface where atmosphere meets land, brings a sense of connectedness and reverence for the abundance and vitality of sustaining Nature.

Principle 5: An Ecocentric Worldview Values Diversity of Ecosystems and Cultures

A major revelation of the Earth-centered perspective is the amazing variety and richness of ecosystems and their organic/inorganic parts. The Earth's surface presents an aesthetically appealing diversity of arctic, temperate and tropical ecosystems. Within this global mosaic the many different varieties of plants, animals, and humans are dependent on their accompanying medley of landforms, soils, waters and local climates. Thus biodiversity, the diversity of organisms, depends on maintenance of ecodiversity, the diversity of ecosystems. Cultural diversity - a form of biodiversity - is the historical result of humans fitting their activities, thoughts and language to specific geographic ecosystems. Therefore, whatever degrades and destroys ecosystems is both a biological and a cultural danger and disgrace. An ecocentric worldview values Earth's diversity in all its forms, the non-human as well as the human.

Each human culture of the past developed a unique language rooted aesthetically and ethically in the sights, sounds, scents, tastes, and feelings of the particular part of Earth that was home to it. Such ecosystem-based cultural diversity was vital, fostering ways of sustainable living in different parts of Earth. Today the ecological languages of aboriginal people, and the cultural diversity they represent, are as endangered as tropical forest species and for the same reasons: the world is being homogenized, ecosystems are being simplified, diversity is declining, variety is being lost. Ecocentric ethics challenges today's economic globalization that ignores the ecological wisdom embedded in diverse cultures, and destroys them for short-term profit.

Principle 6: Ecocentric Ethics Support Social Justice

Many of the injustices within human society hinge on inequality. As such they comprise a subset of the larger injustices and inequities visited by humans on Earth's ecosystems and their species. With its extended forms of community, ecocentrism emphasizes the importance of all interactive components of Earth, including many whose functions are largely unknown. Thus the intrinsic value of all ecosystem parts, organic and inorganic, is established without prohibiting their careful use. "Diversity with Equality" is the standard: an ecological law based on Nature's functioning that provides an ethical guideline for human society.

Social ecologists justly criticize the hierarchical organization within cultures that discriminates against the powerless, especially against disadvantaged women and children. The argument that progress toward sustainable living will be impeded until cultural advancement eases the tensions arising from social injustice and gender inequality, is correct as far as it goes. What it fails to consider is the current rapid degradation of Earth's ecosystems that increases inter-human tensions while foreclosing possibilities for sustainable living and for the elimination of poverty. Social justice issues, however important, cannot be resolved unless the hemorrhaging of ecosystems is stopped by putting an end to homocentric philosophies and activities.

Action Principles

Principle 7: Defend and Preserve Earth's Creative Potential

The originating powers of the Ecosphere are expressed through its resilient geographic ecosystems. Therefore, as first priority, the ecocentric philosophy urges preservation and restoration of natural ecosystems and their component species. Barring planet-destroying collisions with comets and asteroids, Earth's evolving inventiveness will continue for millions of years, hampered only where humans have destroyed whole ecosystems by exterminating species or by toxifying sediments, water and air. The permanent darkness of extinction removes strands in the organic web, reducing the beauty of the Earth and the potential for the future emergence of unique ecosystems with companion organisms, some possibly of greater-than-human sensitivity and intelligence.

"The first rule of intelligent tinkering is to save all the parts" (Aldo Leopold - Sand County Almanac). Actions that unmake the stability and health of the Ecosphere and its ecosystems need to be identified and publicly condemned. Among the most destructive of human activities are militarism and its gross expenditures, the mining of toxic materials, the manufacture of biological poisons in all forms, industrial farming, industrial fishing, and industrial forestry. Unless curbed, lethal technologies such as these, justified as necessary for protecting specific human populations, enriching special corporate interests, and satisfying human wants rather than needs, will lead to ever-greater ecological and social disasters.

Principle 8: Reduce Human Population Size

A primary cause of ecosystem destruction and species extinctions is the burgeoning human population that already far exceeds ecologically sustainable levels. Total world population, now at 6.5 billion, is inexorably climbing by 75 million a year. Every additional human is an environmental "user" on a planet whose capacity to provide for all its creatures is size-limited. In all lands the pressure of numbers continues to undermine the integrity and generative functioning of terrestrial, fresh water, and marine ecosystems. Our human monoculture is overwhelming and destroying Nature's polycultures. Country by country, world population size must be reduced by reducing conceptions.

Ecocentric ethics that value Earth and its evolved systems over species, condemns the social acceptance of unlimited human fecundity. Present need to reduce numbers is greatest in wealthy countries where per capita use of energy and Earth materials is highest. A reasonable objective is the reduction to population levels as they were before the widespread use of fossil fuels; that is, to one billion or less. This will be accomplished either by intelligent policies or inevitably by plague, famine, and warfare.

Principle 9: Reduce Human Consumption of Earth Parts

The chief threat to the Ecosphere's diversity, beauty and stability is the ever-increasing appropriation of the planet's goods for exclusive human uses. Such appropriation and over-use, often justified by population overgrowth, steals the livelihood of other organisms. The selfish homocentric view that humans have the right to all ecosystem components - air, land, water, organisms - is morally reprehensible. Unlike plants, we humans are "heterotrophs" (other-feeders) and must kill to feed, clothe and shelter ourselves, but this is no license to plunder and exterminate. The accelerating consumption of Earth's vital parts is a recipe for destruction of ecodiversity and biodiversity. Wealthy nations armed with powerful technology are the chief offenders, best able to reduce consumption and share with those whose living standards are lowest, but no nation is blameless.

The eternal growth ideology of the market must be renounced, as well as the perverse industrial and economic policies based on it. The Limits to Growth thesis is wise. One rational step toward curbing exploitive economic expansion is the ending of public subsidies to those industries that pollute air, land or water and/or destroy organisms and soils. A philosophy of symbiosis, of living compliantly as a member of Earth's communities, will ensure the restoration of productive ecosystems. For sustainable economies, the guiding beacons are qualitative, not quantitative. "Guard the health, beauty and permanence of land, water, and air, and productivity will look after itself" (E.F. Schumacher - Small is Beautiful).

Principle 10: Promote Ecocentric Governance

Homocentric concepts of governance that encourage over-exploitation and destruction of Earth's ecosystems must be replaced by those beneficial to the survival and integrity of the Ecosphere and its components. Advocates for the vital structures and functions of the Ecosphere are needed as influential members of governing bodies. Such "ecopoliticians," knowledgeable about the processes of Earth and about human ecology, will give voice to the voiceless. In present centers of power, "Who speaks for wolf?" and "Who speaks for temperate rain forest?" Such questions have more than metaphorical significance; they reveal the necessity of legally safeguarding the many essential non-human components of the Ecosphere.

A body of environmental law that confers legal standing on the Ecosphere's vital structures and functions is required. Country by country, ecologically responsible people must be elected or appointed to governing bodies. Appropriate attorney-guardians will act as defendants when ecosystems and their fundamental processes are threatened. Issues will be settled on the basis of preserving ecosystem integrity, not on preserving economic gain. Over time, new bodies of law, policy, and administration will emerge as embodiments of the ecocentric philosophy, ushering in ecocentric methods of governance. Implementation will necessarily be step by slow step over the long term, as people test practical ways to represent and secure the welfare of essential, other-than-human parts of Earth and its ecosystems.

Principle 11: Spread the Message

Those who agree with the preceding principles have a duty to spread the word by education and leadership. The initial urgent task is to awaken all people to their functional dependence on Earth's ecosystems, as well as to their bonds with other species. An outward shift in focus from homocentrism to ecocentrism follows, providing an external ethical regulator for the human enterprise. Such a shift signals what must be done to perpetuate the evolutionary potential of a beautiful Ecosphere. It reveals the necessity of participating in Earth-wise community activities, each playing a personal part in sustaining the marvelous surrounding reality.

This Ecocentric Manifesto is not anti-human, though it rejects chauvinistic homocentrism. By promoting a quest for abiding values - a culture of compliance and symbiosis with this lone Living Planet - it fosters a unifying outlook. The opposite perspective, looking inward without comprehension of the outward, is ever a danger as warring humanistic ideologies, religions, and sects clearly show. Spreading the ecological message, emphasizing humanity's shared outer reality, opens a new and promising path toward international understanding, cooperation, stability and peace.

Acknowledgments for "A Manifesto for Earth"

We thank the following persons for offering critical remarks and commentaries on earlier drafts of this article: Ian Whyte, Jon Legg, Sheila Thomson, Stan Errett, Howard Clifford, Tony Cassils, Marc Saner, Steve Kurtz and Doug Woodard of Ontario; Michelle Church of Manitoba; Don Kerr and Eli Bornstein of Saskatchewan; David Orton of Nova Scotia; Alan Drengson, Bob Barrigar and Robert Harrington of British Columbia; Cathy Ripley of Alberta; Holmes Rolston III of Colorado; David Rothenberg of Massachusetts; Burton Barnes of Michigan; Paul Mosquin of North Carolina; Edward Goldsmith, Patrick Curry and Sandy Irvine of the UK, and Ariel Salleh of Australia. Their helpful reviews do not imply endorsement of this Manifesto for which the authors take full responsibility.

Acknowledgements

We wish to extend sincere appreciation to Dr. Ted Mosquin, and the late Dr. Stan Rowe whose studious labors and cooperation created "A Manifesto for Earth," an insightful and sensible series of principles that should be adopted everywhere. The Manifesto appears in the nick of time because humanity's economic colossus is out of control. One of Stan's dying wishes was that the Manifesto would "get legs," and we sincerely hope that we are instrumental in helping to achieve this by spreading its message. We are grateful for the encouragement of Ted Mosquin and astute comments and recommendations he made regarding our manuscript.

We also wish to thank Don Kerr of NeWest Press in Edmonton for permission to reprint the Manifesto in its entirety as an appendix to this book. NeWest Press is the publisher of a collection of essays by Stan Rowe entitled *Earth Alive.* "A Manifesto for Earth," first published in *BioDiversity: Journal of Life on Earth,* is the final essay in this book. We also thank Don for good suggestions he gave about our work in progress.

Katherine Chomiak, Stan's partner, was also very excited about the project and we thank her for her support and encouragement.

Special thanks to Toshie Sumida, Reiki Master and dedicated humanitarian.

We appreciate Ed Finn for his comments and for sharing many valuable ideas with us over the years, both personally and through his inspired editorship of *The CCPA Monitor,* a monthly journal of The Canadian Centre for Policy Alternatives, which enriches the lives of all who read it.

We are grateful to the late Colleen McCrory of The Valhalla Wilderness Society for her lifelong devotion to the Earth and its future. Her stunning array of environmental achievements are worthy of enormous respect. Colleen is an impressive example of a Canadian woman who dedicated her life to giving testimony for the Earth.

Our thanks to Roland Sattler for proof reading and for his excellent technical help, and for the fine example he gives that one's own life can be guided by a philosophy of moderation and simplicity.

We also appreciate the many insightful people at Hancock House who helped this project become a reality, especially editor Theresa Laviolette, and designer Mia Hancock. We are particularly grateful to Ingrid Luters who came up with the inspiring title, *Testimony for Earth.*

And finally we thank Maggie Oliver for the "just right" drawings that stem from her own love of nature and her immense talent. Our heartfelt appreciation, Maggie, for the outstanding artwork that adorns the cover and pages of this book. We also want to express our gratitude to Roger Oliver who was so helpful with the many technical aspects of the work.

Index

About the Authors

Linda and Bob Harrington live with their bullmastiff, Shadow, on a forested acreage at Galena Bay, British Columbia. They have also reforested a piece of logged-over land by planting thousands of trees and scattering millions of tree seeds. This story is told in *The Soul Solution,* recently reprinted by Hancock House, with a new foreword by David Suzuki.

Bob is a veteran of WW II, and holds a degree in geology and a master's degree in education. He is the author of several books, and numerous feature articles for newspapers and magazines. His diverse experience includes working as a prospector, as a geologist on dam construction, on the USS *Nautilus* during its construction, and teaching secondary school sciences and university ecology courses. He was elected selectman and, later, chairman of the Board of Selectman, in a New England town. He served for several years on the editorial committee of the Journal of the Idaho Academy of Science, and he was Western Representative for the Canadian Wildlife Federation for five years.

Working in the 1950s for the USDA as a foreman–supervisor of pesticide spray operations led him to focus on the study of ecology as the overall holistic science. Fellowships in ecology gave him more insight in this field. In 1990 Bob received the British Columbia Environmental Education Award.

Linda has worked with Bob on his books and articles for many years, and shares his passion for Nature. She loves gardening, collecting and sharing wild and traditional garden seeds, and working to ensure they leave as small a footprint as possible on the planet.

About the Artist

Maggie Oliver trained as an illustrator at the Art Center College of Design in Los Angeles. She now lives in the village of Procter BC, and specializes in large landscape oil paintings of the West Kootenay area of southeastern British Columbia. Maggie can be contacted at mountainlandscapes@gmail.com